Greene Vardiman Black

Descriptive Anatomy of the Human Teeth

Greene Vardiman Black

Descriptive Anatomy of the Human Teeth

ISBN/EAN: 9783337365752

Printed in Europe, USA, Canada, Australia, Japan

Cover: Foto ©berggeist007 / pixelio.de

More available books at **www.hansebooks.com**

Descriptive Anatomy

OF THE

Human Teeth.

BY

G. V. BLACK, M.D., D.D.S.

PUBLISHED BY

THE WILMINGTON DENTAL MANUFACTURING CO.,

1413 FILBERT STREET,

PHILADELPHIA, PA.

PREFACE.

BY my experience as a practitioner, as a teacher, and in my intercourse with fellow practitioners, I have become convinced of a serious defect in the teaching of the details of the anatomy of the teeth, and in the systemization of the terms used in their description. This defect has been a constant drawback at the chair, in the laboratory, and, most of all, in the college. The object of the present volume is to remedy, in a measure, this defect. To this end I have had constantly in view the needs of the dental student and practitioner.

We have heretofore had excellent general descriptions in human and comparative dental anatomy; but these have dealt principally with the general forms of the dentitions of the mammalia and other orders of animate beings, rather than with specific descriptions of the forms of the various surfaces, and surface markings, making up the sum of the forms of the individual teeth of man. Valuable as these works have been, they have left the acquirement of a knowledge of the details of the specific forms of the human teeth mostly to individual observation. By this means, many have attained to an excellent perception of the various forms of the human teeth; but it is not reasonable to suppose the profession generally will do this without some fixed guide. What the dental student wants most in the college, and, in the office, is a systematized nomenclature of the several parts of the teeth in detail; and such a description as will call his attention successively to every part of each tooth, as Gray, in his Anatomy, has called attention to every part of each bone, however apparently unimportant. It should be remembered that anatomy is not to be learned from books alone, but also by bringing the parts to be studied into view, and closely examining them in connection with the descriptions given. Any one who may read the present volume without a reasonable number of human teeth of each denomination before him for examination and comparison, will be but partially benefited.

It has been my object to systematize the nomenclature most in vogue with the profession, whenever practical, rather than to introduce new terms. However, the reader will find a few new terms, and possibly a few old ones that are used differently from the former custom. The terms up and down, to indicate direction or parts of teeth, are abandoned, because of their ambiguity. In a few instances, new forms of old words have been used, especially to avoid the terms inner, outer, backward, forward, etc., which are so often misleading. The words mesial, distal, labial, buccal, lingual, etc., are used as adverbs of direction by adding *ly*, or the same thing is accomplished by the use of the preposition *to*. It is as easy to say of a cavity that it extends far beyond, beyond, to, or nearly to, the gingival line, as to say it extends up or down, etc., and the meaning will not be mistaken; or to say that a cavity extends distally, or to the distal, or lingually, or to the lingual, instead of backward, or inward, either of which have different meanings in different situations. The best rule is to use no extraneous object in the designation of the parts of, or direction on, the surface of a tooth; but to confine the phraseology to the specific and well defined terms applied to its several parts. The back part of a molar would not mean the same relative part as the back part of an incisor. In many such ways the author has endeavored to systematize, and make more definite, the phraseology applied to the teeth without going to extremes, knowing well that forms of language once in use can be improved more easily than they can be displaced by new terms, though more exact.

The absence of a Bibliography may be noted. The plan and object of this work has not seemed to call for many references to authorities. This does not imply, however, that authors who have preceded me, as Fox, Carabelli, Tomes, Wedl, Judd, Wortman, and many others, to whom we are greatly indebted, have been either overlooked or ignored.

The illustrations have all been made by the author for the purpose of illustration, rather than as works of art. After experimenting with the various plans of the management of light and shade, diffuse light has been used because more detail could be shown, especially in the difficult task of illustrating the occluding surfaces of the teeth. Each picture of the teeth, in all its details, is drawn from accurate measurements of the particular tooth in hand.

Much of dental histology might properly find place in this book; but that subject is well represented by others. Malforms of the teeth, supernumerary teeth, and variations of arrangement, belong to the subject of irregularities, which is amply treated by several authors. Our aim has been to confine the book strictly to normal macroscopic anatomy. However, a very serious difficulty, which has always met the dental anatomist, has been the variations of form met in teeth of the same denomination. The endeavor has been to systematize these under one, two, or more typical forms of each tooth, or its lobes, and point to the character of the changes which occur. This has occasionally led to the mention of abnormal forms.

The reader will find scattered through the work some hints with regard to the practical bearing of anatomical points on operative procedures, which it is hoped will be of value.

1. Man's food is both animal and vegetable, and his teeth are so formed as to enable him to readily masticate either kind; therefore, his teeth differ from those of both the carnivorous and herbivorous animals, and form the type of the omnivora. They are formed for cutting, tearing, and comminuting many kinds of food. The incisiors, situated anteriorly, have edges for cutting; the cuspids and bicuspids, at the angles of the mouth, have fairly sharp, though not very long, points or cusps, well calculated for tearing; while the molars, situated in the posterior part of the mouth, have broad, tuberculated grinding surfaces, which serve well to grind or comminute the more solid masses. The forms of the teeth of man indicate a design that his food should be taken in rather small masses, considered from the standpoint of the habits of animals in general, and that it should be very thoroughly commingled with saliva before being passed to the stomach.

2. The adult has thirty-two teeth, as follows: Incisors, $\frac{2}{2}$; cuspids, $\frac{1}{1}$; bicuspids, $\frac{2}{2}$; molars, $\frac{3}{3} = 32$. The teeth are composed of four tissues: enamel, which covers the crown; dentine, which forms the body of both crown and root; cementum, which covers the roots and joins with the enamel at the gingival line, or neck, of the tooth; and pulp tissue, which fills the central cavity of the dentine.

3. Each tooth presents for description certain characters common to all, such as crown, neck or gingival line, root, pulp-chamber, canal, and surfaces of crown and root. The crown of a tooth is that part which projects beyond

7

the gum tissue, and is covered with enamel; while the root, covered with cementum, is that portion fixed in the bony process of the jaw, by which the whole tooth is held securely in position. The root of a tooth may be single, as in the incisors and cuspids; divided into two prongs or roots, as in the lower molars; or into three roots, as in the upper molars; or into a greater number, as in exceptional examples. The root is divided into the body or main portion; the apex, or terminal end of a root; and the neck, which marks the junction of the root with the crown. A tooth is also marked with a slight constriction at the neck, and by the junction of the enamel with the cementum. The latter, which forms a visible line encircling the tooth, is called the gingival line. It is so curved in its course as to present a convexity toward the crown on the proximate surfaces, to correspond with the line of the gum as it passes over the alveolar ridge from the labial to the lingual side of the arch. On the anterior teeth, it is also so curved in passing the labial and lingual surfaces as to present a concavity toward the crown. These are called the curvatures of the gingival line, or the gingival curvature. The neck of a tooth is common to all of the roots, whatever the number, for the point of division into two or more roots is always rootward from the neck.

4.   The crowns of the incisors and cuspids present for examination four surfaces and an edge; and the crowns of the bicuspids and molars, five surfaces. These surfaces are named according to their position and use. Those of the incisors and cuspids presenting toward the lips, are called labial surfaces; those of the bicuspids and molars presenting toward the cheek, buccal surfaces; all presenting toward the tongue, lingual surfaces, whether in the upper or the lower jaw.*

5. The surfaces of the teeth that present toward, or lie against, adjoining teeth are called proximate, or proximal,

---

* Some authors use the term "palatine surfaces" for those of the upper jaw, and "lingual" for those of the lower. This seems unnecessary.

surfaces. The proximate surfaces are also more closely defined by the terms mesial and distal. These terms have special reference to the position of the surface relative to the central or the median line of the face. This line is drawn along the suture uniting the superior maxillary bones, or perpendicularly through the centre of the face and mouth, and passes between the central incisors of both the upper and lower jaws. Those proximate surfaces which, as they are placed in the arch, *and following its curve*, are toward this median line, are called mesial surfaces; and those most distant from this median line are called distal surfaces. The mesial surfaces of the central incisors, both upper and lower, proximate each other; but in all other cases a mesial surface proximates a distal. Also, a distal surface always proximates a mesial surface, except those of the third molars, upper and lower, which have no distal proximating teeth.

6. The angles formed by the junction of any of these surfaces are designated by combining the names of the two uniting surfaces into a compound word, using the term mesio or disto as a prefix, thus : *mesio-buccal, mesio-occluding, disto-lingual* and *disto-labial* angles.

7. The incisors present a cutting edge by the junction of the labial and lingual surfaces along a line. In the cuspids, this joining of the surfaces to form an edge is raised to a point near the center of its length, forming a cusp, hence the term " cuspid," a tooth with one point. The cutting edges of the incisors, and the grinding surfaces of the bicuspids and molars, are those which occlude with the similar surfaces of the teeth of the opposing jaw when the mouth is closed, as in the act of biting ; hence they are called the occluding surfaces. The incisors and cuspids of the upper jaw do not occlude exactly on the cutting edges or cusps, but generally just back of them; though, for convenience, this term is applied to them as if they did. The occluding surfaces of the bicuspids have two cusps, hence the term bicuspid, a tooth with

two points; and the occluding surfaces of the molars have four, and sometimes more.

8. A cusp is a pronounced elevation, more or less pointed, on the surface of a tooth, but more especially on the occluding surface. A slight elevation is often called a tubercle, as that frequently seen near the gingival margin of the lingual surface of the upper incisors. These are generally deviations from the typical forms of the teeth.

9. Long-shaped elevations on the surfaces of teeth are called ridges, and are named according to their location or form: as buccal ridge, lingual ridge, and marginal ridge. Those ridges which descend from the cusps of the molars and bicuspids toward the central part of the occluding surfaces are called triangular ridges. They are named after the cusps to which they belong, as, the triangular ridge of the mesio-buccal cusp of the upper first molar, or simply, mesio-buccal triangular ridge. When a buccal and a lingual triangular ridge join they form a transverse ridge. In this way they often subdivide the central fossa of the lower molars and form supplemental fossæ.

10. A generally rounded or angular depression on the surface of a tooth is called a fossa. Fossæ occur mostly on the occluding surfaces of the molars. When a notable depression is long-shaped it is called a sulcus. Some of these pass entirely from mesial to distal through the occluding surface of a tooth, as in the bicuspids. This term is often erroneously applied to the grooves and fissures.

11. A shallow, long-shaped depression, in the form of a line, on the surface of a tooth, is called a groove. When such a groove follows the bottom of a sulcus it is said to be sulcate. When such a groove sinks suddenly into the substance of a tooth in the form of a fault it is called a fissure.*

---

* The words sulcus, groove, and fissure are often used as if they were synonymous and interchangeable, which has given rise to much confusion. It is necessary that the distinctions made in their use in dental anatomy be well understood. A groove is a very fine line in the form of a slight, sharp depression on the surface of the tooth, as represented in section in diagram A, at *a* in a section cut at the point where the mesial

There are two varieties of grooves in the teeth, differing essentially in their nature and formation. One class marks the lines on which the parts of the teeth first formed separately, are afterward joined, and is called the essential or developmental grooves, or developmental lines. In their relation to the development of the enamel, they are of the same nature, and resemble the sutures of the bones of the skull, which mark the junction of the separately formed plates. Hence these grooves form the key to much of descriptive dental anatomy; and, when they can be traced, enable the anatomist to unravel complex forms, and assign irregularly formed teeth to the groups to which they belong; to properly name their individual parts, and identify any additional, accidental, or unusual forms or parts that may have become intermingled in the accident of a faulty formation. These developmental grooves, or lines, are subject to fissures, which occur when, from any cause, there is a failure of perfect union of the parts, leaving a fault.

12. The other class of grooves has no especial relation to the developmental lines, and are called supplemental grooves. These are aptly described as wrinkles in the enamel, which, in fact, many of them are. But some are so constant in their location and form as to make up a portion of the typical tooth form. The supplemental grooves are usually

groove passes over the marginal ridge of a bicuspid. Diagram B, from a section of a bicuspid, a cut across very close to the occluding surface grooves are shown at a, a. A very shallow depression with rounded bottom is also called a groove, as those seen on the labial surfaces of the incisors. A sulcate groove is one that follows a sulcus of notable depth, the inclines of which approach each other at an angle, as represented at a, in Diagram C. A fissure is always a fault in the enamel, a point where the margins of the plates of enamel fail to unite, as represented at a, in diagram D. These are distinctions of importance, and there should be no confusion in the use of the words by which we distinguish them.

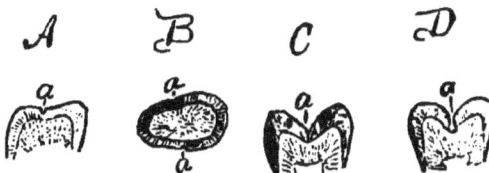

A   *B*   *C*   *D*

shallow, with well-rounded bottoms, and are not ordinarily subject to fissure.

13. In the mesio-distal direction all of the teeth are a little broader at or near their occluding surfaces than at their necks; therefore, as they stand in the well-formed arch, their proximate surfaces touch only at or near their occluding surfaces, leaving **V**-shaped openings between their necks. These are called inter-proximate spaces. Normally, the inter-proximate spaces are filled with gum tissue.

14. The teeth of different individuals show considerable variety of form. Some persons have teeth with very long crowns, broad in the mesio-distal direction at their occluding surfaces and narrow at their necks. These present large inter-proximate spaces. They are known as " bell-crowned " teeth. Again, some individuals have teeth that, in their mesio-distal diameter, are nearly as thick at their necks as at the occluding surfaces, making their inter-proximate spaces very narrow, the teeth almost, or quite, touching along the whole length of the crown. These are known as thick-necked teeth. The more common form is midway between these two extremes. The teeth of some individuals and families have very long cusps; those of others have very short cusps. Some are deeply marked by grooves and sulci, and in those of others the grooves and sulci are shallow. Thus, there is considerable variety of contour without change of type.

15. In the following tables the results of the measurement of many teeth of each denomination are given. The numbers we had of the different varieties differed, but in all denominations they were sufficient to insure reasonable accuracy as to the average size. There are three measurements given, the average, greatest, and least, in the several positions measured. With a greater number, both larger and smaller teeth might be found, so that the tables must not be taken to represent the greatest nor the least that might

be found; but the occurrence of larger, or longer teeth must be rare.

The lines of measurement, are:

1st. " Length over all:" Length of the tooth from the cutting edge, or buccal cusp, to the apex of the root.

2d. " Length of crown:" Length of the crown from the cutting edge, or buccal cusp, to the gingival line on the labial or buccal surface.

3d. "Length of root:" Length of root from the gingival line on the buccal surface to the apex of the root.

4th. "Mesio-distal diameter of crown:" This is the extent from mesial to distal in the greatest diameter, or at the points of proximate contact.

5th. "Mesio-distal diameter of neck:" This measurement was made at the gingival line.

6th. " Labio- or bucco-lingual diameter:" This measurement was taken at the greatest diameter of the crown in the direction named. In the incisors it was on the gingival ridge. In the bicuspids and molars it was generally mid. length of the crown, but occasionally it was near the gingival line, especially in the upper second and third molars.

7th. "Curvature of the gingival line:" This is the height or extent of the curve of the gingival line toward the cutting edge, or occluding surface, as it passes from labial to lingual, measured on the mesial surface.

Having these tables, the necessity for giving many measurements in the text is avoided.

TABLE OF MEASUREMENTS OF THE TEETH OF MAN, GIVEN IN MILLIMETERS AND TENTHS OF MILLIMETERS.*		Length over all.	Length of crown.	Length of root.	Mesio-distal diameter of crown.	Mesio-distal diameter of neck.	Labio- or bucco-lingual diameter.	Curvature of the gingival line.
UPPER TEETH.								
Central Incisor.	Average.	22.5	10.0	12.0	9.0	6.3	7.0	3.0
	Greatest.	27.0	12.0	16.0	10.0	7.0	8.0	4.0
	Least.	18.0	8.0	8.0	8.0	5.5	7.0	2.0
Lateral Incisor.	Average.	22.0	8.8	13.0	6.4	4.4	6.0	2.8
	Greatest.	26.0	10.5	16.0	7.0	5.0	7.0	4.0
	Least.	17.0	8.0	8.0	5.0	4.0	5.0	2.0
Cuspid.	Average.	26.5	9.5	17.3	7.6	5.2	8.0	2.5
	Greatest.	32.0	12.0	20.5	9.0	6.0	9.0	3.5
	Least.	20.0	8.0	11.0	7.0	4.0	7.0	1.0
First Bicuspid.	Average.	20.6	8.2	12.4	7.2	4.9	9.1	1.1
	Greatest.	22.5	9.0	14.0	8.0	6.0	10.0	2.0
	Least.	17.0	7.0	10.0	7.0	4.0	8.0	0.0
Second Bicuspid.	Average.	21.5	7.5	14.0	6.8	5.3	8.8	0.8
	Greatest.	27.0	9.0	19.0	8.0	6.5	10.0	1.5
	Least.	16.0	7.0	10.0	6.0	4.5	7.5	0.0
First Molar.	Average.	20.8	7.7	13.2	10.7	7.5	11.8	2.2
	Greatest.	24.0	9.0	16.0	12.0	8.0	12.0	3.0
	Least.	17.0	7.0	10.0	9.0	7.0	11.0	1.0
Second Molar.	Average.	20.0	7.2	13.0	9.2	6.7	11.5	1.6
	Greatest.	24.0	8.0	17.0	10.0	8.0	12.5	4.0
	Least.	16.0	6.0	9.0	7.0	6.0	10.0	0.0
Third Molar.	Average.	17.1	6.3	11.4	8.6	6.1	10.6	0.7
	Greatest.	22.0	8.0	15.0	11.0	8.0	14.5	2.5
	Least.	14.0	5.0	8.0	7.0	5.0	8.0	0.0

*There are 25.4 millimeters to the inch.

TABLE OF MEASUREMENTS OF THE TEETH OF MAN, GIVEN IN MILLIMETERS AND TENTHS OF MILLIMETERS.		Length over all.	Length of crown.	Length of root.	Mesio-distal diameter of crown.	Mesio-distal diameter of neck.	Labio- or bucco-lingual diameter.	Curvature of the gingival line.
LOWER TEETH.								
Central Incisor.	Average.	20.7	8.8	11.8	5.4	3.5	6.0	2.5
	Greatest.	24.0	10.5	16.0	6.0	5.0	6.5	3.0
	Least.	16.0	7.0	9.0	5.0	2.5	5.5	1.5
Lateral Incisor.	Average.	21.1	9.6	12.7	5.9	3.8	6.4	2.5
	Greatest.	27.0	12.0	17.0	6.5	5 0	7.5	3.5
	Least.	18.0	7.0	11.0	5.0	3.0	6.0	2.0
Cuspid.	Average.	25.6	10.3	15.3	6.9	5 2	7.9	2.9
	Greatest.	32.5	12.0	21.0	9.0	7.0	10.0	4.0
	Least.	20.0	8.0	11.0	5.0	3.0	6.0	2.0
First Bicuspid.	Average.	21.6	7.8	14.0	6.9	4.7	7.7	0.8
	Greatest.	26.0	9.0	18.0	8.0	5.0	8.0	1.5
	Least.	18.0	6.5	11.0	6.0	4.5	7.0	0.5
Second Bicuspid.	Average.	22.3	7.9	14.4	7.1	4.8	8.0	0.6
	Greatest.	26.0	10.0	17.5	8.0	6.5	9.0	2.0
	Least.	18.0	6.0	11.5	6.5	4.0	7.0	0.0
First Molar.	Average.	21.0	7.7	13.2	11.2	8.5	10.3	1.1
	Greatest.	24.0	10.0	15.0	12.0	9.5	11.5	2.0
	Least.	18.0	7.0	11.0	11.0	7.5	10.0	0.0
Second Molar.	Average.	19.8	6.9	12.9	10.7	8.1	10.1	0.2
	Greatest.	22.0	8.0	14.0	11.0	8.5	10.5	1.0
	Least.	18.0	6.0	12.0	10.0	8.0	9.5	0.0
Third Molar.	Average.	18.5	6.7	11.8	10.7	8.3	9.8	0.2
	Greatest.	20.0	8.0	17.0	12.0	9.5	10.5	1.5
	Least.	16.0	6.0	8.0	8.0	5.0	9.0	0.0

C. P. ELA...

TABLE OF MEASUREMENTS OF THE DECIDUOUS TEETH OF MAN, GIVEN IN MILLIMETERS AND TENTHS OF MILLIMETERS. *Averages Only.*	Length over all.	Length of crown.	Length of root.	Mesio-distal diameter of crown.	Mesio-distal diameter of neck.	Labio-lingual diameter of crown.	Labio-lingual diameter of neck.
UPPER TEETH.							
Central Incisor.	16.0	6.0	10.0	6.5	4.5	5.0	4.0
Lateral Incisor.	15.8	5.6	11.4	5.1	3.7	4.8	3.7
Cuspid.	19.0	6.5	13.5	7.0	5.1	7.0	5.5
First Molar.	15.2	5.1	10.0	7.3	5.2	8.5	6.9
Second Molar.	17.5	5.7	11.7	8.2	6.4	10.0	8.3
LOWER TEETH.							
Central Incisors.	14.0	5.0	9.0	4.2	3.0	4.0	3.5
Lateral Incisors.	15.0	5.2	10.0	4.1	3.0	4.0	3.5
Cuspids.	17.0	6.0	11.5	5.0	3.7	4.8	4.0
First Molar.	15.8	6.0	9.8	7.7	6.5	7.0	5.3
Second Molar.	18.8	5.5	11.3	9.9	7.2	8.7	6.4

TABLE OF MEASUREMENTS OF THE TEETH OF MAN; IN INCHES AND HUNDREDTHS OF AN INCH.		Length over all.	Length of crown.	Length of root.	Mesio-distal diameter of crown.	Mesio-distal diameter of neck.	Labio- or bucco-lingual diameter.	Curvature of the gingival line.
UPPER TEETH.								
Central Incisor.	Average.	.88	.39	.49	.35	.24	.27	.11
	Greatest.	1.06	.41	.63	.39	.27	.31	.15
	Least.	.72	.31	.31	.31	.21	.27	.07
Lateral Incisor.	Average.	.86	.34	.51	.25	.17	.23	.11
	Greatest.	1.02	.41	.63	.27	.19	.23	.15
	Least.	.66	.31	.31	.19	.15	.19	.07
Cuspid.	Average.	1.04	.37	.68	.29	.20	.31	.09
	Greatest.	1.26	.47	.80	.35	.23	.35	.13
	Least.	.79	.31	.43	.27	.15	.27	.08
First Bicuspid.	Average.	.81	.32	.48	.28	.19	.35	.04 ·
	Greatest.	.89	.35	.55	.31	.23	.39	.07
	Least.	.66	.27	.39	.27	.15	.31	.00
Second Bicuspid.	Average.	.84	.29	.55	.26	.20	.34	.03
	Greatest.	1.06	.29	.55	.31	.25	.39	.05
	Least.	.62	.27	.39	.23	.17	.29	.00
First Molar.	Average.	.81	.30	.51	.42	.29	.46	.08
	Greatest.	.94	.35	.62	.47	.31	.47	.11
	Least.	.66	.27	.39	.35	.27	.43	.03
Second Molar.	Average.	.78	.28	.51	.36	.26	.45	.05
	Greatest.	.94	.31	.66	.39	.31	.49	.15
	Least.	.62	.23	.35	.27	.23	.39	.00
Third Molar.	Average.	.67	.24	.44	.33	.23	.41	.02
	Greatest.	.86	.31	.59	.43	.31	.57	.09
	Least.	.55	.19	.31	.27	.19	.31	.00

B

TABLE OF MEASUREMENTS OF THE TEETH OF MAN; IN INCHES AND HUNDREDTHS OF AN INCH. LOWER TEETH.		Length over all.	Length of crown.	Length of root.	Mesio-distal diameter of crown.	Mesio-distal diameter of neck.	Labio- or bucco-lingual diameter.	Curvature of the gingival line.
Central Incisor.	Average.	.80	.34	.47	.22	.13	.23	.09
	Greatest.	.94	.41	.62	.23	.19	.26	.11
	Least.	.63	.27	.35	.19	.09	.21	.05
Lateral Incisor.	Average.	.83	.35	.50	.23	.15	.25	.09
	Greatest.	1.06	.46	.66	.26	.19	.29	.13
	Least.	.70	.27	.43	.19	.11	.25	.08
Cuspid.	Average.	1.01	.40	.60	.27	.20	.31	.11
	Greatest.	1.28	.46	.82	.35	.27	.39	.14
	Least.	.78	.32	.43	.23	.11	.24	.08
First Bicuspid.	Average.	.84	.30	.54	.27	.18	.30	.03
	Greatest.	1.02	.35	.70	.32	.20	.31	.05
	Least.	.71	.25	.43	.23	.16	.27	.01
Second Bicuspid.	Average.	.87	.31	.56	.28	.18	.31	.02
	Greatest.	1.02	.43	.63	.32	.25	.35	.07
	Least.	.71	.23	.45	.25	.15	.27	.00
First Molar.	Average.	.78	.30	.52	.33	.33	.40	.04
	Greatest.	.94	.39	.59	.37	.37	.45	.07
	Least.	.71	.27	.43	.23	.29	.39	.00
Second Molar.	Average.	.78	.27	.50	.42	.32	.39	.00
	Greatest.	.86	.31	.55	.43	.33	.41	.00
	Least.	.71	.23	.47	.39	.31	.37	.00
Third Molar.	Average.	.72	.26	.36	.42	.32	.38	.00
	Greatest.	.78	.32	.66	.47	.37	.41	.04
	Least.	.63	.23	.31	.31	.20	.35	.00

TABLE OF MEASUREMENTS OF THE DECIDUOUS TEETH OF MAN, GIVEN IN INCHES AND HUNDREDTHS OF AN INCH. *Averages Only.*	Length over all.	Length of crown.	Length of root.	Mesio-distal diameter of crown.	Mesio-distal diameter of neck.	Labio-lingual diameter of crown.	Labio-lingual diameter of neck.
UPPER TEETH.							
Central Incisor.	.63	.23	.39	.25	.18	.20	.16
Lateral Incisor.	.62	.25	.45	.20	.14	.19	.14
Cuspid.	.74	.25	.53	.27	.20	.27	.21
First Molar.	.59	.20	.39	.28	.20	.33	.27
Second Molar.	.68	.22	.46	.32	.25	.39	.32
LOWER TEETH.							
Central Incisor.	.55	.19	.35	.15	.11	.15	.13
Lateral Incisor.	.59	.19	.39	.15	.11	.15	.13
Cuspid.	.66	.23	.45	.19	.14	.17	.15
First Molar.	.62	.24	.38	.30	.25	.27	.21
Second Molar.	.62	.21	.44	.38	.28	.34	.25

UPPER CENTRAL INCISORS.*

16. The right and left upper central incisors are situated in the extreme anterior part of the dental arch, one on each side of the median line, their mesial surfaces proximating each other. The crown presents four surfaces (five, including the cutting edge), four angles, and a cutting edge, or occluding surface. The general contour of the crown is similar to a wedge, with rounded angles and merging into a rounded form at the thick end, or the neck of the tooth. It

---

* Usually, in the descriptions of the teeth, those of one side only will be mentioned, without reference to which side in the text. Accompanying the illustrations, the side to which the tooth belongs will be given. The student will readily determine to which side a given example belongs by comparing it with the text and correctly naming its surfaces.

is slightly bent on its shorter diameter, so as to make the flattened labial surfaces convex; while the other, the lingual, is concave. The crown is also slightly bent in the mesio-distal direction, so that the labial surface is convex and the lingual concave in this direction. Therefore, there is a general convexity of the labial, and a general concavity of the lingual surface.

17. The labial surface of the crown of the upper central incisor (Fig. 1), in its general form, is an imperfect square, with its gingival side rounded. The mesial margin is a little longer than the distal, so that the cutting edge slopes away toward the distal angle (Fig. 1, h). Both the angles, formed by the proximate surfaces and cutting edge, are slightly rounded, the distal more than the mesial, after which the proximate surfaces converge toward the long axis of the tooth, making the crown a little narrower at the neck than at the cutting edge.

18. The lingual surface of the crown (Fig. 2) is concave in all directions, forming a fossa, bounded by the cutting edge (a), the mesial and distal marginal ridges (n, m) and the gingival ridge, or cingulum (d). The marginal ridges are strong elevations of the enamel, running from the mesial and distal angles of the cutting edge along the borders of this surface to near the gingival line, where they join the gingival ridge. The gingival ridge is a strong elevation of the enamel forming the lingo-gingival border of the crown, sometimes elevated into a tubercle. The lingual fossa is usually smooth, and the ridges by which it is bounded are not prominent. In many instances, however, there is a deep pit at the junction of the gingival ridge with the lingual surface proper; and in some a groove extends from the pit for a short distance along the border of each marginal ridge (Fig. 3). These latter may, or may not, be fissured. In a few examples the enamel of this surface has irregular wrinkles, or ridges and grooves, running from the gingival

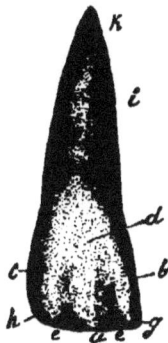

Fig. 1.　　　　　　　Fig. 2.　　　　　　　Fig. 3.

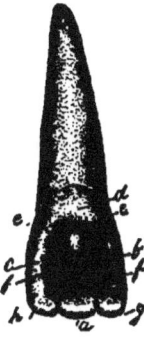

Fig. 4.　　　　　　　Fig. 5.

Fig. 1 * (Par. 17).—RIGHT UPPER CENTRAL INCISOR, LABIAL SURFACE. a, Cutting edge; b, mesial surface; c, distal surface; d, labial surface; e, e, labial grooves; g, mesial angle; h, distal angle; i, body of root; k, apex of root.

Fig. 2 * (Par. 18).—RIGHT UPPER CENTRAL INCISOR, LINGUAL SURFACE. a, Cutting edge; b, mesial surface; c, distal surface; d, gingival ridge, or cingulum; g, mesial angle; h, distal angle; i, body of root; k, apex of root; m, distal marginal ridge; n, mesial marginal ridge.

Fig. 3 * (Par. 18).—LEFT UPPER CENTRAL INCISOR, LINGUAL SURFACE, showing lingual pit. The cutting edge, a, is considerably worn. The mesial and distal marginal ridges, b, c, are prominent; d, linguo-gingival ridge; f, gingival line; g, mesial angle; h, distal angle; i, body of root; k, apex of root; m, lingual pit.

Fig. 4 * (Par. 19).—RIGHT UPPER CENTRAL INCISOR, MESIAL SURFACE. a, Mesial angle; d, gingival ridge; f, f, gingival line, showing its labio-lingual curvature; i, body of root; k, apex of root.

Fig. 5 * (Par. 21).—LEFT UPPER CENTRAL INCISOR, LINGUAL SURFACE. Young, unworn tooth. The developmental lines, or grooves, are made diagrammatically prominent to show the form of the lobes. a, Middle lobe; b, mesial lobe; c distal lobe; d, lingual lobe; e, e, linguo-gingival groove; f, f, lingual developmental grooves; g, mesial angle; h, distal angle.

---

* Illustration, 1½ diameters.

ridge toward the cutting edge. In malformed teeth this surface is often very imperfect.

19. The mesial and distal surfaces each present the outline of the letter V, with its lines curved with the convexity toward the lips, and the acute angle at the cutting edge (Fig. 4). The mesial surface is almost straight from the angle of the cutting edge to the gingival line. It is convex from labial to lingual, but nearly flat toward the gingival line; while in some there is even a slight concavity, centrally at, or near, the gingival line. In the labio-lingual direction, the distal surface is rounded, as in the mesial. In the majority of examples it is also convex in the direction of the long axis of the tooth, so that it bellies out toward the lateral incisor.

20. At the gingival line, the tooth is a little constricted, forming a slight furrow; or rather, the root at the neck is a little smaller than the crown, and the enamel slopes down to the size of the root, giving the appearance of a continuous ridge of enamel around the neck of the tooth. This line does not run horizontally around the neck of the tooth. On the proximate side it forms a curve with the convexity toward the crown, and on the labial and lingual surfaces it forms a curve with the concavity toward the crown. This line marks the limit of the attachment of the peridental membrane to the root of the tooth.

21. *Developmental lines* (Fig. 5). When any of the incisors first appear through the gums there are three little eminences, or tubercles, on the cutting edge with grooves crossing from labial to lingual between them (a,g,h). These grooves run some distance on the labial surface, becoming broader and shallower till they disappear. In many, these lines appear on the lingual surface between the marginal ridges and the fossa (f,f). Occasionally, they are seen as far as the gingival ridge. The little tubercles are soon removed from the edge by wear, leaving it straight, or slightly curved.

These lines divide this part of the crown of the tooth into
three labial lobes. Calcification begins in these tubercles as
separate pieces, or plates, and the grooves are the marks of the
after-confluence of these plates. This is common to the incisors and cuspids.' These teeth are sometimes fissured across the
cutting edge, marking an imperfect confluence of the primary
plates. The calcification of the gingival ridge, or cingulum, is
also begun as a separate plate, forming the lingual lobe, but
afterward it becomes united to the other parts by confluence,
leaving a groove, often very slight, indeed, and soon obliterated by wear, marking the line of union. This is the linguo-
gingival groove (e, e). In smooth regularly formed teeth it
begins at the gingival line just distal to the summit of its
labio-lingual curvature, and runs across the marginal ridge at
right angles with its length, then runs almost horizontally
across the lingual surface to the distal marginal ridge. This
ridge is now crossed at right angles, and the gingival line
reached. The length of the groove usually includes from a
quarter to a third of the circumference of the tooth. When
the gingival ridge is prominent, or rises in the form of a
tubercle, this groove is subject to much variation in its
course. Often, there is a deep pit in the center of its length;
i.e., centrally in the lingual surface, at the margin of the
gingival ridge (Fig. 3, m). From this, fissures may extend
laterally. Occasionally, especially in the lateral incisors, a
sulcus, or a fissure divides the gingival ridge from one of the
marginal ridges, and extends into the cementum. This is
the gingival fissure.

22. The root of the upper central incisor (Figs. 1 to 5)
is about one and a fourth, to one and a half times as long as
the crown. It is conical in form, tapering from the crown
to the apex; less rapidly near the neck, and more rapidly as
the apex is approached. Therefore, the body of the root
seems a little swollen. However, the root of this tooth
presents great variety of figure, as do the roots of teeth

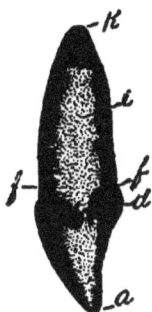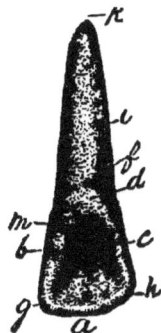

Fig. 6.	Fig. 7.	Fig. 8.	Fig. 9.

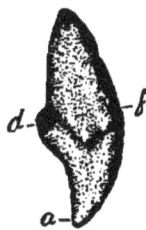

Fig. 10.	Fig. 11.

FIG. 6* (Par. 24).—RIGHT UPPER LATERAL INCISOR, LABIAL SURFACE. *a*, Cutting edge; *c*, distal surface; *e*, labial grooves; *f*, gingival line; *g*, mesial angle; *h*, distal angle.

FIG. 7* (Par. 25).—RIGHT UPPER LATERAL INCISOR, MESIAL SURFACE. *a*, Mesial angle; *d*, linguo-gingival ridge; *f, f*, gingival line; *i*, body of root; *k*, apex of root.

FIG. 8* (Par. 26).—RIGHT UPPER LATERAL INCISOR, LINGUAL SURFACE, without lingual pit. *a*, Cutting edge; *b*, mesial marginal ridge; *c*, distal marginal ridge; *f*, gingival line; *g*, mesial angle; *h*, distal angle; *i*, body of root; *k*, apex of root; *m*, lingual fossa.

FIG. 9* (Par. 26).—RIGHT UPPER LATERAL INCISOR, LINGUAL SURFACE, with lingual pit. *a*, Cutting edge; *b*, mesial marginal ridge; *c*, distal marginal ridge with linguo-gingival groove crossing it; *d*, linguo-gingival ridge, or cingulum; *f*, gingival line; *g*, mesial angle; *h*, distal angle; *i*, body of root; *k*, apex of root; *m*, lingual pit.

FIG. 10* (Par. 26).—RIGHT UPPER LATERAL INCISOR, showing a linguo-gingival fissure. *a*, Cutting edge; *b*, linguo-gingival groove: *f*, gingival line; *g*, mesial angle; *h*, distal angle.

FIG. 11* (Par. 27).—UPPER LATERAL INCISOR, MESIAL SURFACE. Very short root. *a*, Mesial angle; *d*, linguo-gingival ridge; *f*, gingival line.

---

* Illustration, 1½ diameters.

generally. The root is nearly round at the neck. The curvature of the lingual surface is the arc of a smaller circle than that of the labial. The proximate surfaces are slightly flattened. The flattened portion of the mesial surface is a little broader than the distal. These two converge to the lingual, giving the form of a prism with its angles rounded.

UPPER LATERAL INCISOR.

23. The description of the lateral incisor may be much abridged, because of its resemblance to the central in its general form and developmental lines. The tooth is a little shorter, and from mesial to distal the crown is about a third narrower.

24. The labial surface of the lateral incisor (Fig. 6) is more rounded in the mesio-distal direction than in the central. The mesial angle is acute, and the cutting edge slopes away in a curve to a rounded and obtuse distal angle. The cutting edge, at the time of eruption, presents three tubercles, and the grooves crossing the edge between these are projected on the labial surface as shallow labial grooves.

25. The mesial (Fig. 7) and distal surfaces present the characteristic **V**-shape of all the incisors. From labial to lingual the mesial surface is rounded near the cutting edge, but much flattened near the gingival line. Sometimes a slight concavity exists at this point. Occasionally the mesio-labial angle has a flattened or sunken point of enamel near the middle of its length. This is sometimes broad and of notable depth, and in this case is generally in the labial portion of the mesial surface. In others, it is a small imperfection in the mesial border of the labial surface. The distal surface is convex in all directions. In its occlusive third, it rounds out freely toward the cuspid, but becomes more flattened toward the gingival line.

26. The lingual surface (Fig. 8) of lateral incisors, is very irregular in the extent of its concavity. Some are almost flat, while others are deeply concave. The mesial

and distal marginal ridges are proportionately broader, and stronger, than in the centrals. In the majority of examples the lingual surface is the broadest part of the crown. The rounding of the proximate surfaces is at the expense of the labial surface, so that a moderately acute angle is formed by the junction of the proximate surfaces with the lingual. Generally, the lingual surface is almost smooth, but in many, a pit, with or without lateral fissures, will be found at the junction of the lingual surface proper, with the gingival ridge (Fig. 9). In some of these, the gingival ridge is unusually short, so that the marginal ridges are folded in together at their gingival ends, forming a deep sulcus between them, and there is a deep pit at their junction. Again, some are found in which there is a deep groove, which is often fissured, dividing one marginal ridge from the gingival ridge, and extending into the cementum (Fig. 10). This is sometimes nearly central, giving the appearance of a failure of the lingual lobe, or of a division of the lobe centrally, or of the displacement of the lobe to one side. This is the gingival fissure.

27. The root of the upper lateral incisor (Figs. 6 to 11) is conical, but considerably flattened on its mesial and distal sides, which is generally maintained to the apex. The root is generally straight, and about one and a half times as long as the crown. In many specimens the apex is curved to the distal. Occasionally the root is very crooked.

28. The upper lateral incisor presents much variety of size and form. It is not uncommon to find narrow laterals associated with broad centrals. The lateral incisors are more often imperfectly developed than the other anterior teeth. In these, the crown of the tooth is frequently conical, with a rounded, or even a moderately sharp, point.

THE LOWER INCISORS.

29. The lower incisors have outlines similar to the upper lateral, but are, in every way, more slender. Their

Fig. 12.   Fig. 13.   Fig. 14.   Fig. 15.

Fig. 16.   Fig. 17.

FIG. 12 * (Par. 29).—LEFT LOWER CENTRAL INCISOR, LABIAL SURFACE. Long root. a, Cutting edge; e, labial grooves; f, gingival line; g, mesial angle; h, distal angle.

FIG. 13 * (Par. 29).—LEFT LOWER LATERAL INCISOR, LABIAL SURFACE. Long root. a, Cutting edge; e, labial grooves; f, gingival line; g, mesial angle; h, distal angle; i, body of root; k, apex of root.

FIG. 14 * (Par. 29).—RIGHT LOWER LATERAL INCISOR, LABIAL SURFACE. Short root. a, Cutting edge; e, labial grooves; f, gingival line; g, mesial angle; h, distal angle; i, body of root; k, apex of root.

FIG. 15 * (Par. 30).—LOWER CENTRAL INCISOR, LINGUAL SURFACE. a, Cutting edge; b, mesial marginal ridge; c, distal marginal ridge; d, linguo-gingival ridge; g, mesial angle; h, distal angle; m, lingual ridge.

FIG. 16 * (Par. 30).—LOWER CENTRAL INCISOR, DISTAL SURFACE. a, Cutting edge; the edge is worn away as represented by the line; d, linguo-gingival ridge; f, f, gingival line; i, groove along the distal side of root; k, apex of root.

FIG. 17 * (Par. 31).—RIGHT LOWER LATERAL INCISOR, DISTAL SURFACE, showing root deeply grooved. a, Cutting edge; d, linguo-gingival ridge; f, gingival line; i, deep groove in root; k, apex of root.

---

* Illustration, 1½ diameters.

developmental lines are the same, but the grooves are much less marked, and generally cannot be seen except in unworn teeth. The cutting edge of the lower central (Fig. 12) is very nearly at right angles with the long axis of the tooth, and its angles are square and sharp. From mesial to distal, the cutting edge is the widest part of the crown, and from it the proximate surfaces converge equally to the gingival line, reducing the mesio-distal diameter about a third. The lower lateral differs from the central by the cutting edge sloping away to the distal, the mesial angle being acute, and the distal, obtuse and rounded. The distal surface is also convex from the angle to the gingival line, bellying out toward the cuspid.

30. The lingual surface (Fig 15) of the lower incisors is concave from the cutting edge to the gingival ridge, over which there is a convexity (Fig. 16). Near the cutting edge this surface is generally nearly flat in the mesio-distal direction, but is sometimes concave or slightly convex; it becomes convex progressively toward the gingival ridge. In many there is a slight ridge on the center of this surface with a shallow concavity on either side, which marks the junction of the lobes, running from near the cutting edge to the gingival ridge (Fig. 15). The mesial and distal surfaces are convex near the cutting edges, but become flattened, and sometimes slightly concave, toward the gingival line.

31. The roots of the lower incisors are slender and much flattened in their mesio-distal diameter, and not unfrequently slightly grooved on the mesial and distal sides. The labial surface of the crown and root, in its length, forms nearly the arc of a circle (Fig. 17), though the curve of the surface of the crown is usually a little greater than the root (Fig. 16). The lingual surface is almost straight, but in the apical third it is convex, sloping away to form the apex. The roots of these teeth are generally straight, but occasionally the apex is curved to the distal side.

THE CUSPIDS.

32. We have four cuspids, one on each side in the upper and the lower jaw. They are sometimes called canine, or eye, teeth. They are situated at the angles of the mouth between the lateral incisors and the first bicuspids. They are the third tooth from the median line, and are large, strong teeth of simple form, firmly implanted in the alveolar process by a long, strong root, the longest in the human mouth (q. v. tables of measurements). The name cuspid is given to this tooth because its crown is surmounted by a single strong cusp.

THE UPPER CUSPID.

33. The labial surface of the crown of the upper cuspid (Fig. 18) is a little narrower in the mesio-distal direction than in the central incisor, with nearly equal length from the point of the cusp to the gingival line. Instead of a straight, or only slightly curved-cutting edge, as in the incisors, the central portion of the crown is extended into a well-formed point (a), with a cutting edge sloping away to the mesial and distal angles (g, h). Of these cutting edges, the distal is a little the longer, and, from the angle to the gingival line, the distal surface is a little shorter, than the mesial. In unworn teeth, the angle formed by the union of the cutting edges from the cusp is usually about ninety degrees, or a square. The point is a little rounded, but the cusp is soon much rounded or flattened by wear. Both the mesial and distal margins of the labial surface of the crown, from the angles to the gingival line, slope toward the central axis of the tooth,—the distal the most, narrowing the crown of the neck nearly one-third from the width of its widest point. The curvature of the gingival line on this surface, marking the termination of the crown, is about a quarter circle. The surface is convex in all directions, and is much more rounded in the mesio-distal direction, than in the

Fig. 18.   Fig. 19.   Fig. 20.   Fig. 21.

Fig. 22.   Fig. 23.   Fig. 24.

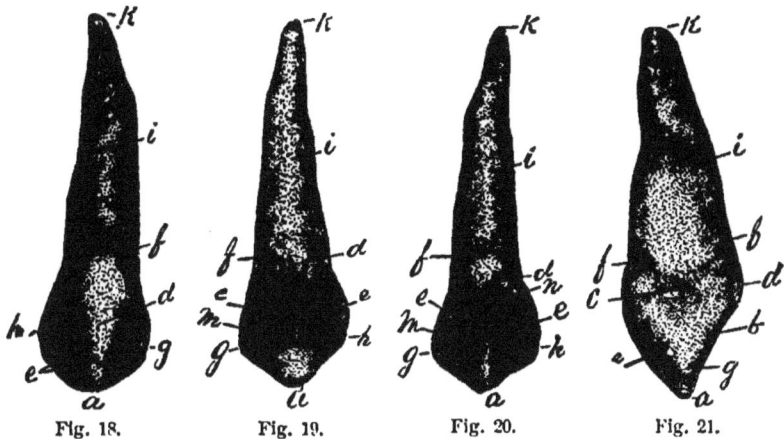

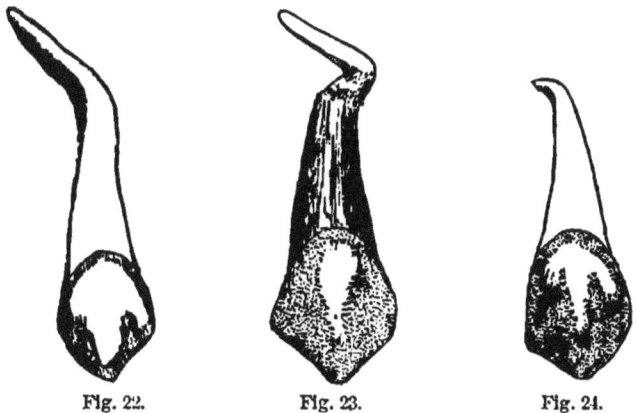

FIG. 18 * (Par. 33).—RIGHT UPPER CUSPID, LABIAL SURFACE.   a, Point of cusp; d, labial ridge; e, labial grooves; f, gingival line; g, mesial angle; h, distal angle; i, body of root; k, apex of root.

FIG. 19 * (Par. 34).—RIGHT UPPER CUSPID, LINGUAL SURFACE.   The more common form. a, Point of cusp; e, e, lingual grooves; d, gingival ridge; f, gingival line; g, mesial angle; h, distal angle; i, body of root; k, apex of root.

FIG. 20 * (Par. 34).—RIGHT UPPER CUSPID, LINGUAL SURFACE of irregular form. a, Point of cusp; d, linguo-gingival ridge; e, e, lingual grooves; f, gingival line; g, mesial angle; h, distal angle; m, prominent lingual ridge; n, two small tubercles on linguo-gingival ridge.

FIG. 21 * (Par. 35).—RIGHT UPPER CUSPID, MESIAL SURFACE.   a, Point of cusp; b, mesial marginal ridge; c, point of slight concavity of mesial surface; d, linguo-gingival ridge; e, labial groove; f, f, gingival line; g, mesial angle; i, body of root; k, apex of root.

FIG. 22 * (Par. 37).—LEFT UPPER CUSPID, with a very long and crooked root.

FIG. 23 * (Par. 37).—RIGHT UPPER CUSPID, with a peculiarly crooked root.

FIG. 24 * (Par. 37).—LEFT UPPER CUSPID, with a very small and short root.

* Illustration, 1½ diameters.

C

incisors. In the direction of its length, the convexity is about the same as in the incisors. The greater convexity mesio-distally is caused by a strong labial ridge (d) running from the point of the cusp to the gingival line. This ridge occupies so much of the surface of the tooth that its margins are imperfectly defined. It belongs to the middle lobe, developed from the middle plate, which, in the incisors, is the smallest of the three, but in this tooth is much the largest. There are two labial furrows (e), or a flattening of the convexity between the central line of the ridge and each angle, marking the junction of the lobes. These furrows are usually lost by becoming shallower before reaching the center of the length of the crown. In well-formed teeth, this surface presents a uniform enamel, free from pits or sulci.

34. The lingual surface (Fig. 19) presents the same general marginal configuration as the buccal, though it is somewhat narrower toward the gingival line. This is caused by the arc of convexity being a smaller circle, and by the flattening of the proximate surfaces on lines which converge rapidly to the lingual. This surface is usually almost straight from the cusp to the gingival ridge, or cingulum, but is sometimes slightly concave. The gingival ridge is sharply convex and longer from the gingival line to the point of convexity than in the incisors. (Fig. 21, d). Mesio-distally this surface is slightly convex in its central part on account of the lingual ridge which runs from the point of the cusp nearly, or quite, to the cingulum. On each side of this, and between it and the marginal ridges, there is a slight but well defined concavity and furrow, marking the confluence of the lobes. The marginal ridges arise from the mesial and distal angles and unite with the gingival ridge or cingulum. These ridges are usually large near the angles, and much less pronounced toward the gingival ridge. The latter is prominent, and is often raised into a tubercle, or slight cusp. Occasionally this part of the enamel is thrown into

irregular folds, with grooves between, which are sometimes fissured. More rarely the small cusp may be divided by a groove (Fig. 20, *n*). The linguo-gingival groove is often pronounced in unworn teeth.

35. The mesial surface, near the angle (Fig. 21), is convex in all directions, but becomes flattened, and occasionally sightly concave, near the gingival line (*c*).

36. The distal surface is similar to the mesial, but is more convex, usually being well rounded in the labio-lingual direction to the gingival line. But in the direction of the long axis of the tooth this surface, on account of the projection of the distal angle, is first convex, and further toward the gingival line is concave, especially near, and at the neck of the tooth. The labio-lingual curvature of the gingival line is about 2.5 m.m., varying from 1.0 m.m. to 3.5 m.m. on the mesial surface, and a little less on the distal.

37. The root of the upper cuspid is the longest in the human mouth, averaging, according to my measurements, 17.5 m.m., and varying from 11.0 to 21.0 from the apex to the gingival line at the buccal surface. It is irregularly conical in form, tapering from the neck to the apex. Its labio-lingual diameter is a little greater than its mesio-distal, which gives the root a flattened appearance, but it is seldom entirely flat on either the mesial or distal surface. In most examples the body of the root is straight, and tapers to a slender point, which is often curved to the labial and distal, though the form of this root presents great variations. Frequently, it is very crooked, perhaps, because, when it is taking its place in the arch, it is often crowded by the teeth mesial and distal to it, so that its growth in a right line is interrupted. (See Figs. 22, 23 and 24).

THE LOWER CUSPIDS.

38. In their general figure, the lower cuspids so closely resemble the upper, a description of their differences will be

Fig. 25.　　　　　　Fig. 26.　　　　　　Fig. 27.

Fig. 28.　　　　　　Fig. 29.

Fig. 25 * (Par. 38).—LEFT LOWER CUSPID, LABIAL SURFACE. *a*, Point of cusp; *b*, mesial surface; *c*, distal surface; *d*, labial ridge; *e*, distal labial groove; *f*, gingival line; *g*, mesial angle; *h*, distal angle; *i*, body of root; *k*, apex of root.

Fig. 26 * (Par. 38).—LEFT LOWER CUSPID, MESIAL SURFACE. *a*, Point of cusp; *d*, linguogingival ridge; *f*, gingival line; *g*, mesial angle; *i*, body of root, which is distinctly flattened; *k*, apex of root.

Fig. 27 * (Par. 38).—LEFT LOWER CUSPID, DISTAL SURFACE. *a*, Point of cusp; *c*, labial groove; *d*, linguo-gingival ridge; *e*, concavity of the distal cutting edge where the groove passes over it; *f*, gingival line; *h*, distal angle; *i*, body of root; *k*, apex of root.

Fig. 28 * (Par. 39).—LEFT LOWER CUSPID, LINGUAL SURFACE. *a*, Point of cusp; *d*, linguo-gingival ridge; *e*, lingual grooves; *f*, gingival line; *g*, mesial angle; *h*, distal angle; *i*, body of root; *k*, apex of root; *l*, distal marginal ridge; *m*, lingual or triangular ridge; *n*, mesial marginal ridge. '

Fig. 29 * (Par. 40).—RIGHT LOWER CUSPID, MESIAL SURFACE. *a*, Point of cusp; *d*, linguo-gingival ridge; *f*, gingival line; *g*, mesial angle. The root is short and thick.

---

* Illustration, 1½ diameters.

sufficient. It is slightly smaller than the upper cuspid, and the crown is a little longer, which makes it appear more slender. The mesial surface is usually nearly straight the entire length of the root and crown (Figs. 25 and 26); so that the increased width of the crown over the root is mainly on the distal. This causes a marked prominence of the distal angle. In many examples this gives the tooth the appearance of being bent, with a considerable concavity on the distal side. In young, unworn teeth, the cusp is rather more prominent and pointed than in the upper cuspid, and the distal cutting edge is proportionately longer; but, as the point of the cusp comes directly in occlusion with the upper teeth, it is soon worn to a blunt point, or a flat surface, inclining to the labial, sloping away to the distal.

39. The lingual surface (Fig. 28) is very smooth, and the ridges are less prominent than in the upper cuspid. A tubercle on the gingival ridge is rare. The developmental lines, or grooves, are the same as those of the upper cuspids, but less prominent. Yet, generally, they can be seen in unworn teeth. Fissures are seldom seen in any part of this tooth.

40. The root of the lower cuspid (Figs. 25 to 29) is shorter than the upper, and generally more flattened in the mesio-distal diameter, often presenting deep furrows. In rare instances, there is a division of the root near the extremity. The root is nearly straight, and in many examples the lingual surface of the root is nearly a straight line, while the buccal surface, root, and crown, present a nearly regular convexity. The root is not so often abnormally crooked as that of the upper cuspid, though, like the upper, the root is much inclined to end in a slender apex; which is often slightly bent in the labial direction.

THE BICUSPIDS.

41. There are eight bicuspids, or premolars, two on each side in the upper jaw, and two on each side in the lower jaw.

Hence they are called the first and second bicuspids. They are situated between the cuspids and first molars, and are the fourth and fifth teeth from the median line. The bicuspids, though unlike the incisors and cuspids in the contour of their crowns, have the same number, and a similar distribution of primary parts, or lobes. They are, indeed, formed on the same general plan. The change of form is the result of a different relative development of the parts, by which the cingulum, or gingival ridge, is elevated into a powerful lingual cusp; which, in the upper bicuspids, is almost or quite as high as the buccal cusp, but in the lower bicuspids, especially in the first, this is less prominent. The central lobe also forms a relatively larger part of the buccal portion of the crown than in the incisors and cuspids; while the mesial and distal lobes are relatively smaller. By the development of the lingual cusp of the upper bicuspids, the linguo-gingival groove of the incisors and cuspids (*q. v.* 21) is carried to the central part of the crown, which it traverses from mesial to distal in a deep sulcus (Fig. 30). It is naturally divided by the mesial and distal pits, found at its junction with the triangular grooves, into three parts: mesial (*o*), central (*l*), and distal (*p*). In the lower bicuspids the lingual lobe is often very small and the course of the grooves irregular.

UPPER FIRST BICUSPID.

42. The outline of the occluding surface of the upper first bicuspid (Fig. 30), when seen in a line with the long axis of the tooth, is irregularly quadrilateral, or trapezoidal in form. The bucco-lingual diameter is about two-ninths greater than the mesio-distal. The flattened proximate surfaces converge toward the lingual, so that the mesio-distal measurement of the buccal portion is a little greater than the lingual. The buccal and lingual surfaces are convex. The buccal convexity forms an arc of about a quarter circle, and merges into the proximate surfaces by obtuse, but well

Fig. 30.   Fig. 31.   Fig. 32.   Fig. 33.

Fig. 34.   Fig. 35.

FIG. 30 * (Par. 42).—RIGHT UPPER FIRST BICUSPID, OCCLUDING SURFACE. *a*, Point of buccal cusp; *b*, lingual cusp; *c*, buccal ridge; *d*, mesial marginal ridge; *e*, distal marginal ridge; *f*, triangular ridge of the buccal cusp; *g*, distal angle; *h*, mesial angle; *i*, triangular ridge of the lingual cusp; *l*, central groove; *o*, mesial groove; *p*, distal groove; *n*, *m*, triangular grooves; *r*, *s*, buccal grooves.

FIG. 31 * (Par. 46).—RIGHT UPPER FIRST BICUSPID, BUCCAL SURFACE. *a*, Buccal cusp; *c*, buccal ridge; *e*, *e*, buccal grooves; *f*, gingival line; *g*, mesial angle; *h*, distal angle; *i*, buccal root; *k*, lingual root.

FIG. 32 * (Par. 48).—RIGHT UPPER FIRST BICUSPID, MESIAL SURFACE. *a*, Buccal cusp; *b*, lingual cusp; *c*, mesial angle; *d*, mesial surface and point where there is often a concavity; *f*, *f*, gingival line; *e*, lingual root; *g*, buccal root.

FIG. 33 * (Par. 50).—UPPER FIRST BICUSPID, with three roots. Bucco-mesial angle.

FIG. 34 * (Par. 50).—UPPER FIRST BICUSPID, with three short roots and of a peculiar form.

FIG. 35 * (Par. 50).—UPPER FIRST BICUSPID, with a single, very long, crooked root.

---

* Illustration, 1½ diameters.

defined, angles (*h*, *g*). The lingual surface forms an arc of nearly half a circle, and merges into the proximate surfaces without any angular prominence.

43. The occluding surface has two prominent cusps— the buccal (*a*), and the lingual (*b*)—and is transversed from mesial to distal by a deep sulcus. The buccal cusp is the larger, and forms the terminal point of the buccal surface. From the point of this cusp, four ridges lead away at right angles. Two of these form cutting edges, which slope away mesially and distally to the mesial (*h*) and distal angles (*g*), where they join the marginal ridges. The central buccal ridge (*c*) leads away centrally on the buccal surface toward the gingival line, forming the convexity of this surface. The triangular ridge (*f*) slopes down to the central part of the crown and joins a similar ridge from the lingual cusp (*i*) to form the transverse ride, or ends in a central sulcate groove (*l*). The lingual cusp (*b*) is in the form of a crescent; its convexity forms the occluding margin of the lingual surface. Instead of a well defined point it usually presents a blunt edge, which runs around its elevated central portion, and joins with the marginal ridges at both angles. The lingual triangular ridge (*i*) leads down from the central point of the cusp to the central groove, to join its fellow from the buccal cusp in the formation of the transverse ridge, or is divided from it by a deep central sulcate groove. The ridge is seldom prominent. Very frequently the central incline of the lingual cusp is a plain surface.

44. The marginal ridges, mesial (*d*) and the distal (*e*), are strong ridges of enamel which rise in the mesial and distal terminations of the cutting edges of the buccal cusp, and form the mesial and distal margins of the occluding surface. They join with the ridge forming the lingual cusp. Or they are usually divided from the latter by the mesial and distal grooves; though the latter are often indistinct, especially in teeth that have been somewhat worn.

45. The occluding surface of the bicuspid has five developmental grooves; the central (l), mesial (o), distal (p), mesial triangular (m), and distal triangular (n). The central groove is deeply sulcate, and divides the triangular ridges, or passes over their junction as a shallow line, and sinks into a triangular pit at either end. The mesial and distal grooves are really continuations of the central, which pass over the marginal ridges as very fine lines, or as more marked grooves, and mark the boundary of the lingual lobe. They are rarely fissured, while the central groove is frequently fissured throughout its course. The triangular grooves, mesial (m) and distal (n), run from the mesial and distal pits toward the mesial and distal angles, dividing the marginal ridges from the triangular. They are occasionally sulcate in the first part of their course, and are generally lost toward the buccal angles by becoming shallower; but in young, unworn teeth they can often be followed as a fine line running over the cutting edges of the buccal cusp near the angles, and leading into the buccal grooves (r, s). These are the marks of confluence of the mesial and distal lobes with the median or central lobe. In the central incline of the lingual cusp, supplemental grooves are often seen meeting the triangular grooves of the buccal side. The triangular grooves are occasionally fissured for a short distance from their junction with the central.

46. The buccal surface of the upper first bicuspid (Fig. 31) is similar to the labial surface of the cuspid ($q. v$, 33). The cusp is usually nearer the center of the crown, and generally somewhat to the distal; therefore, the cutting edges which run from the summit of the cusp to either angle may be of about equal length. In some examples the distal edge is the longer; but usually the mesial edge is the longer. In the gingival half of its length this surface is smoothly convex from mesial to distal; but further toward the occluding margin, the buccal ridge, which terminates in

the cusp, becomes more prominent, and a shallow buccal groove (*e, e*) appears at both sides of the ridge, or between it and the angles. This surface is also considerably narrowed toward the gingival border, almost equally on the mesial and distal, so that the crown seems much broader at the occluding surface.

47. The lingual surface is regularly convex from mesial to distal. From the gingival margin to the summit of the lingual cusp it is often a straight line; but more generally it is slightly convex, in many examples almost as convex as the buccal surface.

48. The mesial surface (Fig. 32) is much flattened from buccal to lingual, but is generally slightly convex over its whole extent; yet in many examples there is a slight concavity near the gingival line. In the direction from the gingival line to the occluding margin, this surface is slightly convex through its whole length, but not equally so through its buccal and lingual half. The lingual portion is progressively more rounded toward the occluding surface; while the buccal portion is nearly straight to the angle.*

49. The form of the distal surface agrees substantially with the mesial, but is rather more convex in all directions and any concavity is rare.

50. The root of the upper first bicuspid is usually either much flattened and grooved on its mesial and distal sides, or separated into two divisions, one-third to two-thirds of its length, making one buccal and one lingual root. More than half have their roots thus divided. When separated, the roots taper regularly to slender apexes. When not divided, the apex is apt to be obtuse. Occasionally this tooth presents three divisions of the root, two buccal and one lingual

---

* The form of the proximate surfaces is especially important in making contour fillings, for any concavity increases the difficulty of forming good, clean margins at the gingival border. The unequal convexity of the buccal and lingual halves is important, and requires a special adaptation of instruments to make a perfect contour and good clean margins.

(Figs. 33, 34). In some instances the root of this tooth is very crooked or otherwise distorted (Fig. 35).

UPPER SECOND BICUSPID.

51. The upper second bicuspid so nearly resembles the upper first, just described, that a notice of its differences will be sufficient. It is a little smaller, and in every way more slender.* The general form of the occluding surface (Fig. 36) is similar to the first bicuspid. It presents a buccal and a lingual cusp, and similar sulcus, ridges, grooves, and pits. The average height of the cusps is considerably less than in the first bicuspid. The marginal ridges are proportionally broader, the mesial and distal pits closer to each other, and the central groove shorter. The triangular grooves join the central groove nearer the mesio-distal center of the tooth, making the buccal triangular ridge narrower and more nearly pointed. In many examples the enamel of the occluding surface is thrown into several shallow wrinkles, or supplemental grooves and ridges, which radiate from the central groove, which occurs but rarely in the first bicuspid. The buccal cusp is a little nearer the mesial than the distal angle, so that the distal edge is slightly the longer.

52. From mesial to distal, the buccal surface (Fig. 37) is not so broad at the occluding surface, and is a little broader at the neck, so that it has not so much of the bell-crown appearance as the first bicuspid. Otherwise, this surface has the same, but less definitely defined, outlines and markings. The mesial and distal surfaces(Figs. 38 and 39) are generally slightly more convex, and the crown more smoothly rounded. The mesial surface seldom shows a concavity. The lingual surface is usually a little more rounded toward the cutting edge or crest of the cusp. Generally, the distal side of the lingual cusp is rounded to such an extent as to bring the summit of the cusp to the mesial of the central line of the

---

* The popular opinion is that the second bicuspid is the larger.

Fig. 36.	Fig. 37.	Fig. 38.	Fig. 39.

Fig. 40.	Fig. 41.

FIG. 36 * (Par. 51).—RIGHT UPPER SECOND BICUSPID, OCCLUDING SURFACE. *a*, Point of buccal cusp; *b*, lingual cusp; *c*, buccal ridge; *e*, mesial marginal ridge; *d*, distal marginal ridge; *f*, triangular ridge of the buccal cusp; *g*, mesial angle; *h*, distal angle; *i*, triangular ridge of the lingual cusp; *l*, central groove; *m*, *n*, triangular grooves; *o*, *p*, buccal grooves.

FIG. 37 * (Par. 52).—RIGHT UPPER SECOND BICUSPID, BUCCAL SURFACE. *a*, Point of buccal cusp; *c*, buccal ridge; *e*, *e*, buccal grooves; *f*, gingival line; *g*, mesial angle; *h*, distal angle; *i*, body of root; *k*, apex of root.

FIG. 38 * (Par. 52).—RIGHT UPPER SECOND BICUSPID, MESIAL SURFACE. *a*, Buccal cusp; *b*, lingual cusp; *f*, gingival line; *i*, groove in the mesial side of the root.

FIG. 39 * (Par. 52).—RIGHT UPPER SECOND BICUSPID, DISTAL SURFACE. *a*, Buccal cusp; *b*, lingual cusp; *c*, carous cavity near the proximate contact point; *f*, gingival line; *i*, groove in distal side of root.

FIG. 40 * (Par. 53).—UPPER SECOND BICUSPID, with a very crooked root.

FIG. 41 * (Par. 53).—UPPER SECOND BICUSPID with a very short crook of the root.

—————————————————————————

* Illustration, 1½ diameters.

tooth. The gingival line, in its course round the neck of the tooth, makes but a slight labio-lingual curvature on the mesial surface. Generally there is no curvature on the distal surface.

53. The root of the upper second bicuspid is a little longer than the first, while the crown is slightly shorter, which makes the proportionate increase of length appear considerable. The root is rarely divided in any part of its length, but is much flattened from the neck to the apex. The mesial side is often deeply grooved in the apical third of its length ; the distal side is less frequently grooved. The root tapers very gradually, remaining broad in the bucco-lingual diameter, and ends in a blunt apex. A few have a root that tapers rapidly, becomes more rounded, and ends in a slender apex. Crooked roots are more frequent in this than in the other bicuspids. (Figs. 40 and 41).

LOWER FIRST BICUSPID.

54. This tooth is the smallest of the bicuspids. The occluding surface differs much from the upper first bicuspid. Indeed, the lingual cusp is so nearly wanting, it would hardly be called a bicuspid except for its association. The buccal cusp is large and prominent, and so much is the buccal surface inclined toward the long axis of the tooth, that, when seen in a line with the long axis of the tooth, its point occupies a position about one-third distant from the buccal toward the lingual outline of the crown (Figs. 42 and 43). The buccal cusp presents the same ridges leading from its summit as described for the upper first bicuspid (43), but the pair which form the cutting edges usually form a curve with its convexity to the buccal, and merge into the marginal ridges by more rounded angles. The buccal triangular ridge is narrow and prominent (*f*), and joins the elevated lingual ridge, or cusp (*b*), forming a complete transverse ridge. In many instances, this is deflected to the mesial or distal. In

D

young teeth, the central groove often crosses the transverse ridge as a fine line, which soon disappears by wear. However, in many examples the transverse ridge is divided by a deeply sulcate groove. There is a deep pit at the mesial and the distal ends of the central groove, or on either side of the transverse ridge (c, d) from which the triangular grooves, which divide the marginal ridges from the triangular, or transverse ridge, run toward the distal and buccal angles. These are often sulcate in the first part of their course. In many young teeth these grooves may be traced over the cutting edges onto the buccal surface, marking the confluence of the central with the mesial and distal buccal lobes.

55. The lingual lobe varies much in its size and outline. It is divided from the three buccal lobes by the mesial, distal, and central grooves; and often occupies but a small portion of the lingual margin of the occluding surface. The grooves which mark its outlines are often indistinct, and often obliterated by wear. The lobe may form a ridge of even height, and join the marginal ridges, or it may be elevated into one or more tubercles. It may, also, be raised into a small cusp, located centrally, or to either side of the central line of the tooth.

56. The marginal ridges are occasionally quite small, but in most examples they are well developed, and cause the mesial and distal surfaces to stand out prominently, giving the tooth a strong bell-crowned appearance. The length of the marginal ridges vary with the size of the lingual lobe.

57. The buccal surface (Fig. 44) is convex in all directions. The cusp is to the distal of its perpendicular line, and its figure similar to the upper first bicuspid, except that its surface is more convex. As the occlusion is directly on the point of the cusp, as with the anterior lower teeth generally, it is soon so worn down that its prominence is lost.

58. The mesial and distal surfaces are convex from buccal to lingual. In the direction from the occluding

Fig. 42.

Fig. 43.

Fig. 44.

Fig. 45.

Fig. 46.

Fig. 47.

Fig. 42 * (Par. 54).—Right Lower First Bicuspid, Occluding Surface. *a*, Point of buccal cusp; *b*, lingual cusp or ridge; *c*, buccal ridge; *d*, mesial marginal ridge; *e*, distal marginal ridge; *f*, triangular ridge of buccal cusp or buccal triangular ridge; *g*, mesial angle; *h*, distal angle; *i*, central groove crossing the transverse ridge; *l*, mesial pit; *o*, *p*, buccal grooves.

Fig. 43 * (Par. 54).—Lower First Bicuspid, Occluding Surface. *a*, Point of buccal cusp; *b*, lingual cusp or ridge; *c*, triangular ridge of buccal cusp deflected to one side; *d*, triangular groove with fissure; *e*, mesial marginal ridge; *f*, distal marginal ridge.

Fig. 44 * (Par. 57).—Right Lower First Bicuspid, Buccal Surface. *a*, Buccal cusp; *d*, buccal ridge; *e*, *e*, buccal grooves; *f*, gingival line; *g*. distal angle; *h*, mesial angle; *i*, body of root; *k*, apex of root.

Fig. 45 * (Par. 58).—Left Lower First Bicuspid, Mesial Surface. *a*, Buccal cusp; *b*, lingual cusp or ridge; *c*, distal marginal ridge; *d*, triangular ridge of buccal cusp; *e*, mesial marginal ridge; *f*, gingival line.

Fig. 46 * (Par. 59).—Left Lower First Bicuspid, Lingual Surface. *a*, Buccal cusp; *b*, lingual cusp or ridge; *c*, distal marginal ridge; *d*, triangular ridge of buccal cusp; *e*, mesial marginal ridge; *f*, gingival line.

Fig. 47 * (Par. 59).—Right Lower First Bicuspid, Lingual Surface. *a*, Buccal cusp; *b*, lingual cusp or ridge; *d*, triangular ridge of buccal cusp; *e*, *e*, marginal ridges; *f*, gingival line.

---

* Illustration, 1½ diameters.

margin to the gingival line, they are generally concave, after passing the convexity of the immediate occluding margin. This latter stands out boldly to both mesial and distal, giving a marked bell-crowned form. The concavity of the mesial and distal surfaces is not so well seen from the buccal view as from the lingual. The greater over-hang of the crown (over the root), and the greater concavity, is toward the lingual portion.

59. The lingual surface (Figs. 46 and 47) is smoothly rounded from mesial to distal, and slightly convex in the direction of the length of the tooth. It is only about half as long as the buccal surface (in unworn teeth), and when viewed at right angles with the long axis of the tooth, all of the occluding surface on the lingual side of the buccal cusp is seen except the pits and sulci.

60. The neck of the tooth is much constricted, and has a deep gingival line at the junction of the enamel and cementum. This makes the proper adjustment of a band for an artificial crown particularly difficult. The curvature of the gingival line is much less than in the upper first bicuspid, the average being less than one millimeter (q. v. table of measurements).

61. The root of this tooth is somewhat flattened at the neck, on lines that converge rapidly toward the lingual, and often grooved on its mesial and distal sides, and sometimes the root is bifurcated. It tapers regularly, with a tendency to a more nearly round form at the apex, and generally ends in a slender point. The root is generally straight, or the lingual surface is straight, and the buccal surface convex. This convexity extends from the apex of the root to the point of the crown, which gives that characteristic appearance so peculiar to the anterior lower teeth.

LOWER SECOND BICUSPID.

62. The lower second bicuspid is a little longer than the lower first, and of much the same figure, except that the

lingual cusp is proportionately higher and more nearly, but never quite, on a level with the buccal, but the lingual surface is about equal to the buccal in mesio-distal breadth. These teeth are regular in general contour, but the grooves of the occluding surface are much diversified. These differences may be classified under three forms: 1st. The central groove joins the triangular grooves in such a way as to form a half circle with the convexity to the lingual, with or without a transverse ridge crossing its line. In these, when the transverse ridge is high, only a deep pit appears on either side (Fig. 48). 2d. The lingual cusp is divided by a sulcate groove, which runs over centrally, or nearly so, to the lingual surface, making a three-cusped tooth (Fig. 49). The central groove forms an angle at the junction with the lingual, or is crescentic in form. It joins with the triangular grooves in such a way that the point of junction cannot be told except by finding the mesial and distal grooves, which are often very indistinct. 3d. The central groove is straight, and generally sulcate, with a deep pit at both ends. In many examples these pits are crossed by the triangular grooves almost at right angles with the central (Fig. 50). By tracing the fine mesial and distal grooves carefully in young, unworn, teeth, it will be found that the lingual lobe is much larger than in the lower first bicuspid, and almost as large as in the upper bicuspids. In the three-cusped forms the two lingual lobes are usually a little broader from mesial to distal than the buccal portion of the tooth.

63. The buccal surface (Fig. 51) of the lower second bicuspid does not differ from the other bicuspids, except in being shorter, the cusp lower, and the lingual surface (Fig. 52) broader and smoothly convex; in the three-cusped forms it is often somewhat flattened and grooved in its occluding third, giving the crown a squarish appearance, or even triangular when the lingual lobes are large.

64. The mesial and distal surfaces in these bicuspids are

Fig. 48.

Fig. 49.

Fig. 50.

Fig. 51.

Fig. 52.

Fig. 53.

FIG. 48 * (Par. 62).—LEFT LOWER SECOND BICUSPID, OCCLUDING SURFACE. *a*, Buccal cusp ; *b*, lingual cusp or ridge ; *c, e*, marginal ridges ; *d,f*, pits ; *g, h*, triangular grooves.

FIG. 49 * (Par. 62).—RIGHT LOWER SECOND BICUSPID, OCCLUDING SURFACE, with three cusps. *a*, Buccal cusp ; *b*, disto-lingual cusp ; *c*, mesio-lingual cusp ; *d*, lingual groove ; *e, e*, mesial and distal grooves.

FIG. 50 * (Par. 62).—RIGHT LOWER SECOND BICUSPID, OCCLUDING SURFACE, with straight central groove. *a*, Buccal cusp ; *b*, lingual cusp ; *c, e*, marginal ridges ; *d*, triangular ridge of the buccal cusp ; *f*, central groove ; *g, h*, triangular grooves.

FIG. 51 * (Par. 63).—LEFT LOWER SECOND BICUSPID, BUCCAL SURFACE. *a*, Buccal cusp ; *d*, buccal ridge ; *e, e*, buccal grooves ; *f*, gingival line ; *g*, mesial angle ; *h*, distal angle ; *i*, body of root ; *k*, apex of root.

FIG. 52 * (Par. 63).—RIGHT LOWER SECOND BICUSPID, LINGUAL SURFACE. *a*, Buccal cusp ; *b*, lingual cusp ; *c*, distal marginal ridge ; *d*, triangular ridge of the buccal cusp ; *e*, mesial marginal ridge ; *f*, gingival line.

FIG. 53 * (Par. 64).—LEFT LOWER SECOND BICUSPID, MESIAL SURFACE. *a*, Buccal cusp ; *b*, lingual cusp ; *d*, triangular ridge ; *e*, mesial marginal ridge ; *f*, gingival line ; *i*, groove in mesial side of the root.

---

* Illustration, 1½ diameters.

a little flattened, but remain convex from buccal to lingual. From the occluding margin to the gingival line, they are nearly straight, though some are convex and some in part concave.

65. The root of the lower second bicuspid is larger and longer than in the first bicuspid. It is flattened on the mesial and distal surfaces on nearly parallel lines. In some examples they are concave or grooved. In rare instances the root of this tooth is grooved on the buccal and lingual sides, with a tendency to a division into a mesial and distal prong. In the greater number it tapers regularly to a slender apex, but in many, the apex is large and obtuse. The root is generally straight, but occasionally very crooked.

THE MOLAR TEETH.

66. The molar teeth are very different in form and plan of construction from those previously described. They are particularly designed for grinding or comminuting food; for this purpose have broad, occluding surfaces, broken by ridges, grooves, and fossæ. The ridges are raised at intervals into powerful cusps, which fit with more or less accuracy into the sulci and fossæ of the opposing teeth. There are twelve molars, three on each side of both jaws. They are the sixth, seventh and eighth teeth from the median line, and are commonly named the first, second and third molars, upper and lower. The last is also called the wisdom tooth, or *Dens Sapientia*. The upper and lower molars are much alike in size and general contour, but in the detail of the arrangement of their lobes, cusps, fossæ and grooves, they are different. It will, therefore, be necessary to describe the upper and lower molars separately.

THE UPPER MOLARS.

67. The three upper molars are similar, but present minor differences of detail, consisting mostly in a less pro-

nounced, or typical development of certain parts, or lobes, of the second and third molars. The upper first molar being the typical form, will be described first, and afterward the deviations from this type that occur in the second and third molars.

UPPER FIRST MOLAR.

68. The occluding surface of the upper first molar (Fig. 54), when seen in a line with the long axis of the tooth, presents an outline of irregular rhombic form, with the mesio-buccal and disto-lingual as acute angles. The angles are rounded, with more or less convexity of the marginal lines. This surface presents two principal fossæ, and four developmental grooves. These grooves divide the crown into four lobes, or primary developmental parts, each of which is surmounted by a strong cusp. These lobes, or cusps, are the mesio-buccal (Fig. 54, a), disto-buccal (b), mesio-lingual (c), and disto-lingual (d). Of the grooves which outline these parts, three rise from the central pit of the central fossa: The mesial (h), which runs to the mesial margin; the buccal (i), which runs in a deep sulcus to the buccal margin, and over it onto the buccal surface; and the distal (j), which runs distally, and lingually, over the transverse ridge and ends in the distal fossa (k). The remaining groove—the disto-lingual (k, k)—begins a little to the buccal of the central part of the distal margin, and runs diagonally in a straight line, or in a curve with the concavity toward the disto-lingual angle, to the lingual margin, and over it onto the lingual surface, to become the lingual groove. Except that part crossing the distal marginal ridge, this groove is usually deeply sulcate.

69. In their origin, each distinct portion begins its calcification as a separate piece, plate, or cusp, and moves apart from the others as growth proceeds, till the proper dimensions of the occluding surface is attained; then they coalesce on the lines marked by these grooves, and thus com-

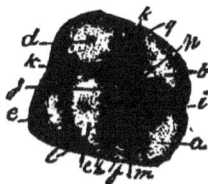

Fig. 54.   Fig. 55.   Fig. 56.

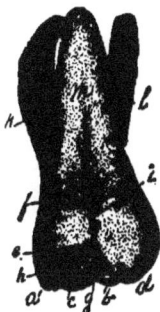
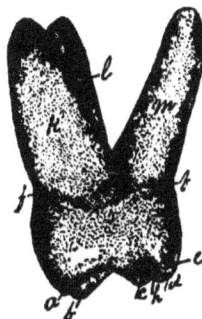

Fig. 57.   Fig. 58.

FIG. 54 * (Par. 68.)—RIGHT UPPER FIRST MOLAR, OCCLUDING SURFACE, with four cusps. *a*, Mesio-buccal cusp; *b*, disto-buccal cusp; *c*, mesio-lingual cusp; *d*, disto-lingual cusp; *f*, mesial marginal ridge; *g*, distal marginal ridge; *h*, mesial groove; *i*, buccal groove; *j*, distal groove; *k, k*, disto-lingual groove; *m*, mesio-buccal triangular groove; *n*, disto-buccal triangular ridge; this unites with the distal ridge from the lingual cusp to form the oblique ridge; *o*, disto-buccal triangular groove; *p*, mesio-buccal triangular ridge; *q*, central pit.

FIG. 55 * (Par. 70).—RIGHT UPPER FIRST MOLAR, OCCLUDING SURFACE, with five cusps. *a*, Mesio-buccal cusp; *b*, disto-buccal cusp; *c*, mesio-lingual cusp; *d*, disto-lingual cusp; *e*, fifth cusp; *f*, mesial marginal ridge; *g*, distal marginal ridge; *h*, mesial groove; *i*, buccal groove; *j*, distal groove; *k, k*, disto-lingual groove; *l, k*, mesio-lingual groove. The lingual cusps are faceted by wear; *m*, mesio-buccal triangular groove; *n*, disto-buccal triangular ridge.

FIG. 56 * (Par. 78).—RIGHT UPPER FIRST MOLAR, BUCCAL SURFACE. *a*, Mesio-buccal cusp; *b*, disto-buccal cusp; *c*, mesio-lingual cusp; *d*, disto-lingual cusp; *e*, buccal ridge; *f*, gingival line; *g*, mesial angle; *h*, distal angle; *i*, buccal groove; *k*, mesial root; *l*, distal root; *m*, lingual root.

FIG. 57 * (Par. 79).—RIGHT UPPER FIRST MOLAR, LINGUAL SURFACE. *a*, Mesio-buccal cusp; *b*, disto-buccal cusp; *c*, mesio-lingual cusp; *d*, disto-lingual cusp; *e*, fifth cusp; *f*, gingival line; *g*, disto-lingual groove; *h*, mesio-lingual groove; *i*, lingual groove; *k*, mesial root; *l*, distal root, *m*, lingual root.

FIG. 58 * (Par. 80).—RIGHT UPPER FIRST MOLAR, MESIAL SURFACE. *a*, Mesio-buccal cusp; *b*, disto-buccal cusp; *c*, mesio-lingual cusp; *d*, disto-lingual cusp; *e*, fifth cusp; *f, f*, gingival line; *h*, mesio-lingual groove; *k*, mesial root; *l*, distal root; *m*, lingual root.

---

* Illustration, 1½ diameters.

plete the occluding surface. When the union has been completed, nothing but fine lines remain, which, on any plain parts, are often quickly obliterated by wear. But on parts meeting at an angle, forming a sulcus, as in the buccal and disto-lingual grooves, there is usually a sharp groove, and at any point where the union has been imperfect, there is a fissure. This fissure is most frequent at the ends of the grooves, near the central pit, or where they are deeply sulcate, and in the central portion of the disto-lingual grooves; but they may occur in any part of these lines.

70. In many examples of the upper first molars, but in no others, there is a small fifth lobe or cusp (Fig. 55, e). This is situated on the lingual side of the mesio-lingual lobe, from which it is divided by a fifth groove, the mesio-lingual (l, k), which runs from the lingual portion of the mesial margin diagonally to the lingual margin, and joins the lingual groove. This cusp, when it occurs, is always bilateral, i.e., on both the right and left upper first molars. It is hereditary, appearing regularly in the teeth of children when present in the teeth of both parents. It occurs also, in a modified form, when present in but one parent. Therefore, the cusp will be found in all possible varieties of development, from its largest size, as represented in Fig. 55, to the merest line marking its position on the lingual side of the mesio-lingual cusp.

71. The occluding surface of the upper first molar has four marginal ridges, broken by the grooves described (68), so as to form four principal eminences or cusps. These are the buccal, lingual, mesial, and distal-marginal ridges. The buccal-marginal ridge begins at the mesio-buccal angle, in the form of a blunt cutting edge, and rises in a curved line to the summit of the mesio-buccal cusp (Fig. 54, a), from which it descends distally to the buccal groove (i). From the summit of this cusp the mesio-buccal triangular ridge (p) descends to the mesial side of the central pit (q). This ridge

is divided from the mesial-marginal ridge by the mesio-buccal supplemental groove (*m*), which is sometimes deep, but generally shallow, and occasionally absent. From the buccal groove the marginal ridge rises rapidly to the summit of the disto-buccal cusp (*b*), then descends in a curve to the disto-buccal angle, to join the distal-marginal ridge (*g*). From the point of this cusp the disto-buccal triangular ridge (or triangular ridge of the disto-buccal cusp, *n*), runs down to the distal side of the central pit, where it joins a ridge from the mesio-lingual cusp to form the oblique ridge, or is divided from this ridge by a sulcate distal groove.

72. The lingual-marginal ridge begins at the mesio-lingual angle as a rounded edge, and rises in a curve to the summit of the mesio-lingual cusp (*c*), and descends, continuing its curve, into a ridge, meeting the triangular ridge of the disto-buccal cusp to form the oblique ridge. However, in a less pronounced form, this marginal ridge descends from the summit of the mesio-lingual cusp to the distal, where it is deeply broken by the disto-lingual groove. On the distal side of this groove it rises abruptly to the summit of the disto-lingual cusp (*d*), from which it slopes away in a curve to join the distal-marginal ridge.

73. The mesial-marginal ridge (*f*) is a strong band of enamel running from the mesio-buccal to the mesio-lingual angle of the tooth. It forms the mesial boundary of the occluding surface, and the angle of junction of the mesial and occluding surfaces, or the mesio-occluding angle. It is low in the center of its length, and rises toward both angles. It is crossed near its center by the mesial groove, usually as a fine line, which is often obliterated by wear early in life. Occasionally one or more small tubercles appear in the central portion of this ridge, inclosed in what seems to be a division of the mesial groove.

74. The distal-marginal ridge (*g*) is a band of enamel forming the distal boundary of the occluding surface, from

angle to angle, and forms the angle of junction between the
occluding and the distal surfaces, or the disto-occluding
angle. It is low in the centre of its course, and is crossed
by the distal end of the disto-lingual groove as a fine line,
usually a little to the buccal of the central point.

75. The central fossa is irregularly circular, and is
formed by the central inclines of the mesial marginal ridge,
mesio-buccal cusp, disto-buccal cusp, mesio-lingual cusp, and
oblique ridge. It is made irregular, and sometimes angular,
by the depth of the several sulcate grooves, or by the prominence of the triangular ridges, especially that from the
mesio-buccal cusp. The central incline of the mesio-lingual
cusp is generally a plain surface, but is sometimes slightly
concave, and in about a fourth of its examples there is a low,
triangular ridge running from its apex directly to the central pit. The oblique ridge is occasionally cut through by a
deep and wide distal sulcus, thus connecting the central with
the distal fossa. In a few examples two supplemental grooves
or wrinkles arise from the mesial groove at about the center
of its length, and run, one toward the mesio-buccal, and the
other toward the mesio-lingual angle of the tooth. These
are often of such depth and width as to form a small supplemental mesial fossa. Of these grooves, the mesio-lingual
is generally absent, and the mesio-buccal (m) is generally
present, though often very shallow.

76. The distal fossa is formed by the distal incline of
the disto-buccal and mesio-lingual cusps and oblique ridge,
which meet the central incline of the disto-lingual cusp and
distal-marginal ridge. It is traversed by the disto-lingual
groove, which is generally sulcate, and deepens into a pit
at the point where it receives the distal groove. From the
buccal portion a supplemental groove, the disto-buccal, runs
well up toward the summit of the disto-buccal cusp, dividing its triangular ridge from the distal-marginal ridge.

77. In some molars many supplemental grooves or

wrinkles radiate from the centers of these fossæ on the inclines of the ridges and cusps, and in some poorly developed teeth these are deeply fissured. The growth of enamel seems to have reached out toward the line of junction in the form of spiculæ, and to have failed to fill out the space.

78. The buccal surface of the upper first molar (Fig. 56) is irregularly convex. Its length is about equal to the mesio-distal breadth at the gingival line, while the width at the widest point, near the occluding margin, is about three-tenths greater. Therefore, the mesial and distal margins converge toward the neck. The mesial margin is almost straight, after the rounding of its angle, but the distal is convex. The occluding margin is surmounted by the buccal cusps (a, b), between which there is a deep notch, through which the buccal groove passes from the occluding to the buccal surface. This groove passes centrally toward the gingival line about half the length of this surface, dividing the occluding portion into a mesial and a distal buccal ridge. In some examples the groove continues across the gingival line to the bifurcation of the roots. There is a bucco-gingival ridge of enamel (e) near the gingival line which the buccal groove generally does not cross. The mesio-buccal convexity is greatest on this ridge, and diminishes toward the occluding margin. The summit of this convexity is on the mesial half, in a line with the mesial root, and the surface falls away rapidly toward the distal.

79. The lingual surface (Fig. 57) is divided in a line with the long axis of the tooth into a mesial and distal lobe by the lingual groove (i). Both lobes are smoothly convex from mesial to distal, and in a less degree from the gingival line (f) to the occluding margin. The gingival line is nearly horizontal, and so sunken as to give the appearance of a gingival enamel ridge. The occluding margin is surmounted by the mesio- and disto-lingual cusps (c, d), of which the mesial is usually the larger. In the five-cusped molar the

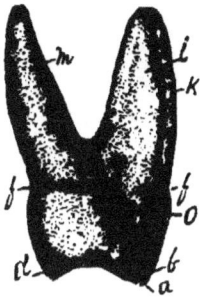
Fig. 59.

Fig. 60.

Fig. 61.

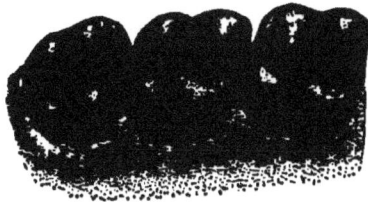
Fig. 62.

FIG. 59 * (Par. 81).—RIGHT UPPER FIRST MOLAR, DISTAL SURFACE. *a*, Mesio-buccal cusp; *b*, disto-buccal cusp; *d*, disto-lingual cusp; *f, f*, gingival line; *k*, mesial root; *l*, distal root; *m*, lingual root; *o*, concavity of the distal surface.

FIG. 60 * (Par. 82).—LEFT UPPER FIRST MOLAR, BUCCAL SURFACE. A bell-crowned tooth.

FIG. 61 * (Par. 85).—UPPER FIRST MOLAR showing the mesial and distal roots united at their apexes.

FIG. 62 * (Par. 86).—THE RIGHT UPPER MOLARS showing the progressive diminution of the disto-lingual cusp from the first to the third molar.

---

* Illustration, 1½ diameters.

E

fifth cusp (*e*) is seen on the mesio-lingual lobe. The mesial
and distal margins are convex, converging rapidly toward
the lingual root. In the few examples in which the lingual
groove is not apparent, the lingual surface is smoothly convex from mesial to distal.

80. The mesial surface (Fig. 58) is nearly flat in all
directions, and its margins are rounded to the buccal and
lingual surfaces. Toward the mesio-buccal angle, the flat
surface is continued well up to the occluding surface, which
it meets in a fairly sharp angle; but as it approaches the
lingual portion it is progressively rounded toward the occluding surface. In many molars, near the gingival line,
half way from mesial to distal, this surface is a little concave.

81. The distal surface (Fig. 59), in its lingual half, is
convex in all directions, but in its buccal half there is often
a concavity formed by a considerable distal protrusion of the
disto-lingual lobe. This is a shallow, but marked, depression
(*o*), running from the bifurcation of the distal and the lingual
roots toward the disto-buccal cusp. It crosses the gingival
line at a point about one-third distant from the buccal toward the distal margin. In many examples this depression
is more central, and receives the distal termination of the
disto-lingual groove, which may often be traced as a fine line
nearly or quite to the gingival border of the enamel. This
complication of the surface makes the finishing of fillings,
and the fitting of bands for crowns, specially difficult. This
depression falls short of forming a concavity in about one-
fourth of the first molars.

82. The root of the upper first molar (Figs. 56 to 61)
is divided into three prongs*—the mesial, or mesio-buccal

---

* The divisions of the root of a tooth are usually called "the roots," and
the upper molars are said to have three roots. The neck of the tooth is,
however, common to all, and, more strictly, there is one root divided into
several prongs, or fangs. Some authors follow this form of expression. For
instance, Prof. Judd says: "The root of a molar is divided into three fangs."
I see no good reason, however, for not calling them roots, the term generally
in use by the dental profession.

(k), the distal or disto-buccal (l), and the lingual (m). These are usually quite widely separated, giving the tooth a firm seat in its alveolus. The lingual root is the largest; it diverges boldly to the lingual, and is straight, or slightly curved with the convexity to the lingual. It is conical, and tapers regularly to an obtuse rounded apex. In most molars it has a groove, a continuation of the lingual, on its lingual side, running nearly, or quite, its whole length. In rare examples this root bifurcates.

83. The mesial root (k) is larger than the distal; broad from labial to lingual, and thin from mesial to distal, with the lingual portion the thinner, and grooved on its flattened sides. It diverges first to the mesial and buccal, and then curves to the distal. It tapers mostly at the expense of the lingual thin edge, and ends in a flattened or rounded apex.

84. The distal root (l) is the smallest of the three. It diverges to the distal and buccal, and is straight, or slightly curved, sometimes to the distal, sometimes to the mesial, so that its apex approaches the mesial root. It is flattened on its mesial and distal sides, but less than the mesial root, and tapers to a more rounded form, ending in a slender point.

85. The roots of the first molar are, perhaps, more regular in form than those of any other of the molar teeth; but even in these, considerable deviation from the forms described will occasionally be observed (Figs. 60 and 61).

UPPER SECOND MOLAR.

86. The most constant difference between the occluding surfaces of the upper first and second molars is that the latter tooth is smaller, and has an almost constant tendency to a relatively smaller size of the disto-lingual lobe. This tendency is well seen in Fig. 62, in which the upper molars of a very well formed denture are drawn from the cast. This shows the disto-lingual lobe progressively diminished, so that in the third molar it is very small. When a large

Fig. 63.   Fig. 64.   Fig. 65.

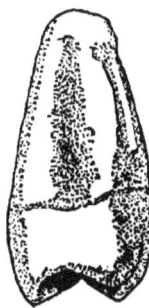

Fig. 66.   Fig. 67.   Fig. 68.

FIG. 63 * (Par. 86).—RIGHT UPPER SECOND MOLAR, OCCLUDING SURFACE, with three of the cusps faceted by wear. Tooth large and especially well developed. *a*, Mesio-buccal cusp; *b*, disto-buccal cusp; *c*, mesio-lingual cusp; *d*, disto-lingual cusp; *f*, mesial marginal ridge; *g*, distal marginal ridge; *h*, mesial groove; *i*, buccal groove; *j*, distal groove; *k*, disto-lingual groove; *l*, distal fossa; *m*, mesio-buccal triangular groove; *n*. central pit; *o*, mesio-buccal triangular ridge; *p*, disto-buccal triangular ridge; *r*, mesio-lingual triangular ridge,

FIG. 64 * (Par. 86).—RIGHT UPPER SECOND MOLAR, OCCLUDING SURFACE. Tooth of medium size and imperfect form. *a*, Mesio-buccal cusp; *b*, disto-buccal-cusp; *c*, mesio-lingual cusp; *d*, disto-lingual cusp; *f*, mesial marginal ridge; *g*, distal marginal ridge; *h*, mesial groove; *i*, buccal groove; *l*, distal fossa. The disto-lingual groove is imperfect; *m*, mesio-buccal triangular groove; *o*, mesio-buccal triangular ridge; *p*, disto-buccal triangular ridge, which unites with the ridge from the mesio-lingual cusp to form the oblique ridge.

FIG. 65 * (Par. 87)—RIGHT UPPER SECOND MOLAR showing a form of deformity peculiar to this tooth that is not very common.

FIG. 66 * (Par. 89).—RIGHT UPPER SECOND MOLAR with the roots inclined to the distal.

FIG. 67 * (Par. 90).—RIGHT UPPER SECOND MOLAR, DISTAL SURFACE, with the three roots compressed into one.

FIG. 68 * (Par. 90).—RIGHT UPPER SECOND MOLAR, showing the distal and lingual roots united.

---

* Illustration, 1½ diameters.

number of second molars are examined, it is found that the central fossa is not so regularly formed as in the first, being often small, even in large, well-developed teeth (Fig. 63), and the cusps comparatively low, with a relatively greater inclination of the lingual surface toward the summit of the mesio-lingual cusp. In many examples the distal fossa is reduced to a mere pit (Fig. 64), and the lingual marginal ridge is continued from the summit of the mesio-lingual cusp distally to the disto-lingual angle, where it joins the distal marginal ridge, as in Fig. 64, making practically a three-cusped tooth. In many of these, after the tooth is a little worn, the disto-lingual groove cannot be traced across the lingual marginal ridge, nor on the lingual surface; but a fine line is generally seen crossing the distal marginal ridge.

87. The upper second molar is liable to a peculiar deformity, which I have not seen in any other tooth. The crown is greatly flattened from mesial to distal, occasionally to such an extent as to present none of its usual lines. In the extreme cases (Fig. 65) there is one central, long-shaped sulcus running from buccal to lingual, formed by the central inclines of a ridge encircling the occluding surface. This ridge may be broken at intervals by shallow grooves crossing it. Usually, much of the central inclination of this ridge is wrinkled, and often there are many small fissures. The mesial surface is commonly deeply concave, and the distal convex.

88. In well-developed second molars, the form of the buccal, lingual, mesial, and distal surfaces are much the same as in the first molar, though rather more convex, and concavities in the mesial and distal surfaces are not so frequent. However, in teeth with large distal cusps, the concavity of the distal surface is about the same as in the first molar (81). The lingual groove is generally not sulcate, though it is usually seen as a fine line in unworn teeth. It is often near the disto-lingual angle, and, when

sulcate, it generally disappears by becoming shallower about half-way toward the gingival line ; rarely, it runs across the gingival line, and in that case the lingual root is grooved, as in the first molars.

89. The neck of the tooth is less regular in its outline than in the first molar. In the average, it is more flattened from mesial to distal, and on lines that converge more to the lingual. On the buccal surface, the mesio-buccal ridge is relatively more prominent, and at the neck the surface slopes away more toward the distal, so that, in many examples, the distal root seems compressed between the mesial and lingual roots.

90. The roots of this tooth are the same in number and general form as in the first molar; but they spread less, and are curved more to the distal (Fig. 66). There is also much more variety of the comparative size of the root and crown than in the first molar. In many the root is single, with the outlines of the division marked by grooves of variable depth (Fig. 67); or two of the roots may be connected, while the third is free. Sometimes this union is the lingual with the distal (Fig. 68); but oftener it is the lingual with the mesial root.

UPPER THIRD MOLAR.

91. The upper third molar deviates from the typical form of the first more than does the second. Yet, in dentures of the best form, it presents the same developmental lines, fossæ, and cusps (Fig. 69). The disto-lingual lobe is very much smaller, and in many examples is entirely wanting. Of the examples in my possession, about twenty-one per cent. show a diminutive disto-lingual cusp; thirty-two per cent. show a pit in the position of the distal fossa, and some portion of the disto-lingual groove, but no depressed groove over the lingual-marginal ridge; it is similar to that shown in Fig. 64. The remainder, or about forty-seven per cent., have no disto-lingual lobe. Of the latter, three-fourths

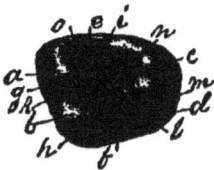
Fig. 69.

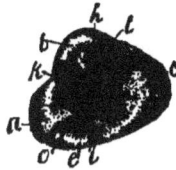
Fig. 70.

Fig. 71.

Fig. 72.

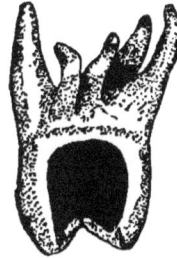
Fig. 73.

Fig. 69* (Par. 91).—RIGHT UPPER THIRD MOLAR, OCCLUDING SURFACE, of typical form. *a*, Mesio-buccal cusp; *b*, disto-buccal cusp; *c*, mesio-lingual cusp; *d*, disto-lingual cusp; *e*, mesial marginal ridge; *f*, distal marginal ridge; *g*, mesio-buccal triangular ridge; *h*, disto-buccal triangular ridge; *i*, mesial groove; *k*, buccal groove; *l*, distal groove; *m* disto-lingual groove; *n*, central pit; *o*, mesio-buccal triangular groove.

Fig. 70* (Par. 91).—RIGHT UPPER THIRD MOLAR, OCCLUDING SURFACE, three cusped. *a*, Mesio-buccal cusp; *b*, disto-buccal cusp; *c*, lingual cusp; *e*, mesial marginal ridge; *h*, distal marginal ridge; *i*, mesial groove; *k*, buccal groove; *l*, distal groove; *o*, mesio-buccal triangular groove.

Fig. 71* (Par. 91).—RIGHT UPPER THIRD MOLAR, OCCLUDING SURFACE. Young, unworn tooth of imperfect form. *a*, Mesio-buccal cusp; *b*, disto-buccal cusp; *c*, lingual cusp; *e*, mesial marginal ridge; *f*, distal marginal ridge; *g*, mesial groove; *h*, buccal groove; *i*, distal groove. Several supplemental grooves are seen radiating from the central pit.

Fig. 72* (Par. 94).—RIGHT UPPER THIRD MOLAR, BUCCAL SURFACE. The deeply grooved single root is not quite complete, and shows the funnel-shaped opening at the apex.

Fig. 73* (Par. 94).—LEFT UPPER THIRD MOLAR with five roots.

---

* Illustration, 1½ diameters.

are properly three-cusped teeth in which the oblique ridge becomes the distal marginal ridge, and the distal groove runs over to the distal surface (Fig. 70). In the remainder, there is only an irregular ridge (Fig. 71), forming a central fossa, so marked with wrinkles or supplemental grooves that the developmental lines are not satisfactorily made out. Many examples of the three-cusped teeth show much relative diminution of the disto-buccal lobe.

92. The upper third molar is the smallest of the molars,* and is more irregular in its size and conformation than the first or the second molar. On account of the small size of the disto-lingual lobe, the distal portion of the tooth is much smaller than the mesial, and in the three-cusped teeth the crown becomes triangular, with its angles well rounded.

93. The mesial surface resembles that in the upper first and second molars, but is more rounded, so that a concave portion is less frequent, and the distal surface is well rounded. The lingual and buccal surfaces are more rounded than the same surfaces in the other upper molars. The gingival line is usually horizontal in its course around the neck of the tooth, except that, in a few examples, there is a slight curvature on the mesial surface. Many upper third molars are much flattened from mesial to distal. These generally have a considerable concavity on the mesial surface. Occasionally teeth are found that are also abnormally small and have the appearance of supernumeraries. More rarely this tooth fails to develop, and is wanting entirely.

94. The root of the upper third molar has, in the more regular forms, the three divisions common to the upper molars, though relatively smaller and not so widely separated as those of the first and second molars; and often end in slender conical points; many have but a single root; but in most of

* Forty upper first molars balanced sixty upper third molars.

these the three roots are outlined by grooves of more or less depth (Fig. 72). A considerable number have more than three roots, which are irregular in size or form (Fig. 73). Four, five, six, and even seven or eight divisions are sometimes met. The neck is then generally broad from buccal to lingual, and the occluding surface of the crown irregularly formed. Many of the three cusped teeth have the root without divisions.

THE LOWER MOLARS.

95. The lower molars differ so much from each other, especially the first and second, that each must be separately described.

THE LOWER FIRST MOLAR.

96. The lower first molar is the sixth tooth from the median line in the lower jaw. It proximates the lower second bicuspid with its mesial surface, and the lower second molar with its distal. Next to the upper first molar, it is the largest tooth in the denture.* The outline of the occluding surface (Figs. 74 and 75), when seen in a line with the long axis of the tooth, is trapezodial, with the buccal line the longer. The buccal angles are about equally acute, while the lingual angles are equally obtuse, and all are more or less rounded. The buccal margin is convex, but made irregular by two buccal grooves. The lingual angle is nearly straight, but sometimes slightly concave, or notched in the center of its length, by the lingual groove; but more generally it is slightly convex. The mesial and the distal lines are nearly straight in the best formed teeth; though the distal is sometimes considerably convex, as the fifth, or disto-lingual cusp is more or less prominent. All of these lines vary very much as to their convexity; the rule being that, in teeth of large size and symmetrical development, they approach nearer to straight lines.

---

* Forty-two upper first molars balanced forty-six lower first molars.

Fig. 74.

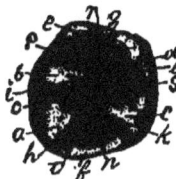

Fig. 75.

Fig. 76.

Fig. 77.

FIG. 74 * (Par. 96).—LEFT LOWER FIRST MOLAR, OCCLUDING SURFACE, of typical form. a, Mesio-buccal cusp; b, disto-buccal cusp; c, mesio-lingual cusp; d, disto-lingual cusp; e, distal cusp; f, mesial marginal ridge; g, distal marginal ridge; h, mesio-buccal triangular ridge; i, disto-buccal triangular ridge; k, disto-lingual triangular ridge; l, mesio-lingual triangular ridge; m, distal triangular ridge; n, mesial groove; o, buccal groove; p, disto-buccal groove; r, distal groove; s, lingual groove.

FIG. 75 * (Par. 96).—RIGHT LOWER FIRST MOLAR, OCCLUDING SURFACE. a, Mesio-buccal cusp; b, disto-buccal cusp; c, mesial lingual cusp; d, disto-lingual cusp; e, distal cusp; f, mesial marginal ridge; g, distal marginal ridge; h, mesial triangular ridge; i, disto-buccal triangular ridge; k, mesio-lingual triangular ridge; l, disto-lingual triangular ridge; n, mesial groove; o, buccal groove; p, disto-buccal groove; r, distal groove; s, lingual groove; t, mesio-buccal triangular groove.

FIG. 76 * (Par. 104).—LEFT LOWER FIRST MOLAR, BUCCAL SURFACE. a, Mesio-buccal cusp; b, disto-buccal cusp; c, distal cusp; d, bucco-gingival ridge; e, buccal pit; f, gingival line; g, mesial root; h, distal root; i, buccal groove; k, disto-buccal groove.

FIG. 77 * (Par. 104).—LEFT LOWER FIRST MOLAR, BUCCAL SURFACE, with prominent cusps. References the same as for Fig. 76.

---

* Illustration, 1½ diameters.

97. Each of these marginal portions is surmounted by the mesial, buccal, lingual, and distal marginal ridges, which form a continuous elevation of irregular height around the margins of the occluding surface, and on which there are five cusps. The central inclinations of these ridges forms the central fossa. On the mesial, lingual, and distal, the summits of these ridges are closed on the margins of the surface; but on the buccal, there is an inward inclination of the buccal surface that carries the summit of the ridge considerably toward the central axis of the tooth.

98. The occluding surface has five developmental grooves (Figs. 74 and 75)—the mesial, buccal, disto-buccal, lingual, and distal—which divide it into five developmental parts, or lobes. These are the mesio-buccal (*a*), disto-buccal (*b*), mesio-lingual (*c*), disto-lingual (*d*), and distal (*e*) lobes; each bearing a cusp of the same name. The mesial grove (*n*) runs from the central fossa over the mesial marginal ridge to the mesial surface. On the mesial marginal ridge it is usually a fine line which is soon obliterated by wear. Occasionally, this is divided into two branches, with a small tubercle on the mesial marginal ridge between them (Fig. 74, *f*). In many examples there is a supplemental groove which rises from the mesial groove at about the center of its length, and runs toward the mesio-buccal angle. This is the mesio-buccal triangular groove (Fig. 75, *t*). It divides the mesial marginal ridge from the triangular ridge of the mesio-buccal cusp. More rarely there is also a similar groove running toward the mesio-lingual cusp. When these are deep, they form a mesial supplemental fossa (Fig. 75, *t*). The buccal groove (*o*) runs in a deep sulcus from the central pit to, and over, the buccal marginal ridge to the buccal surface, and divides the mesio-buccal from the disto-buccal cusp. The disto-buccal groove (*p*) also runs from the central pit over the buccal ridge, more or less near the distal angle, as the distal cusp is large or small. It divides the disto-buccal lobe

from the distal. The lingual grooves (s) runs from the central pit in a deep sulcus to, and over, the lingual marginal ridge onto the lingual surface, and divides the two lingual lobes. The distal groove (r) runs distally over the distal marginal ridge, and divides the disto-lingual lobe from the distal. Frequently this groove can be traced some distance toward the gingival line on the distal surface. The mesial and distal grooves form a line traversing the whole extent of the occluding surface, from mesial to distal, in the center of which a **V**-shaped deflection is formed with its point to the lingual, the base receiving the point of the triangular ridge (i) of the disto-buccal cusp.

99. In most examples, the central fossa occupies all the occluding surface within the circle of the summit of the marginal ridges, though, occasionally, one or more supplemental fossæ are divided from it by high triangular ridges running down from the cusps (Fig. 75, h, k). The surface of the fossa is made irregular in most of these teeth by deep sulci on the lines of the grooves, separating the cusps and triangular ridges.

100. The occluding surface of the lower first molar has five cusps, one on each of the five lobes, or three on the buccal marginal, and two on the lingual marginal ridge.* These cusps are formed by the grooves previously described (98), which pass over the ridges in depressions of variable depth, thus subdividing the crests of the ridges into obtuse elevations. Usually, these are not so high and prominent as the cusps of the upper molars. The mesio-buccal (Figs. 74 and 75, a) is the largest and strongest of the buccal cusps, and occupies rather more than one-third of the buccal marginal ridge. From its crest a triangular ridge (h) runs down centrally to the junction of the mesial and buccal grooves, and is divided from a similar triangular ridge from the mesio-lingual cusp,

---

* In some rare cases lower first molars have but four cusps, and then the tooth is like the lower second molar (109, 110).

by the mesial groove. When these two ridges are high, they form, in conjunction with triangular grooves between them and the mesial marginal ridge, a mesial supplemental fossa. The disto-buccal cusp (*b*) is of less extent from mesial to distal, but has a longer triangular ridge, though not so high, which ends in the point of the **V**-shaped deflection of the mesial and distal grooves, or at their junction.

101. The lingual cusps (*c*, *d*) are about equal in size and height (perhaps the mesial is a little the higher on the average). Each has strong triangular ridges (*k*, *l*) which terminate in the angles formed by the junction of the lingual groove with the mesial and distal grooves in the central pit.

102. The distal cusp (*e*) occupies the distal portion of the buccal ridge, and forms the disto-buccal angle of the tooth. It is the distinguishing mark of the lower first molar, being but very rarely absent in that tooth, and never present in the lower second molar. It is the smallest of the five cusps, and varies most in its relative size. In some examples it is almost or quite as large as the disto-buccal cusp. In others, especially in small and poorly developed teeth, it may be reduced to a mere tubercle, occupying the buccal portion of the distal marginal ridge and the immediate disto-buccal angle. The triangular ridge, or central incline, of this cusp is commonly nearly flat, but occasionally has a rounded crest. It ends in a point at the junction of the distal and disto-buccal grooves.

103. In this tooth there is often a deep pit at the junction of the mesial, distal, and lingual grooves. Less frequently there is also a pit at the junction of the mesial and buccal grooves, and at the junction of the distal and disto-buccal grooves. The grooves are often fissured for a short distance from the pits, especially in the deeply sulcate lingual groove. In some very poorly developed teeth fissures may be found in any part of the grooves.

104. The buccal surface of the lower first molar (Figs.

F

76 and 77), when seen at right angles with the long axis of the tooth, is irregularly trapezoid in form, with the occluding margin about two-sevenths longer than the gingival. The mesial and distal margins converge toward the gingival, and their angles, with the occluding surface, are about equally acute. The occluding margin is broken into three elevations, or cusps, by the buccal and disto-buccal grooves. The gingival line is straight, or slightly curved, with the concavity toward the occluding surface. The mesial and the distal lines are slightly convex. The buccal surface is convex in all directions; but the line of convexity from mesial to distal is broken toward the occluding margin by the buccal and disto-buccal grooves, which pass over from the occluding surface. The buccal groove (i) is usually a little to the mesial of the central line of the surface, and often ends in a deep pit (e), about half way from the occluding to the gingival margin. Exceptionally, this groove is continued to the bifurcation of the root. The disto-buccal groove (k) is near the disto-buccal angle, and its course is toward the gingival line, with a distal inclination. It is usually lost to sight by becoming shallower, but in some examples it may be traced to the gingival line. The enamel terminates in a marked inclination to the gingival line, forming the bucco-gingival ridge.

105. The lingual surface of this tooth (Figs. 78 and 79) is slightly convex in all directions. It forms a fairly sharp angle with the occluding surface, but is rounded away toward the mesial and distal surfaces. On account of the lingual convergence of the mesial and distal surfaces, the lingual surface is much shorter from the mesial to distal than the buccal. The occluding margin is deeply notched by the passage of the lingual groove (i), which usually terminates near the center of the surface by becoming shallower.

106. The mesial surface (Fig. 80) is very irregular in outline, and often the occluding margin is deeply concave.

Fig. 78.

Fig. 79.

Fig. 80.

Fig. 81.

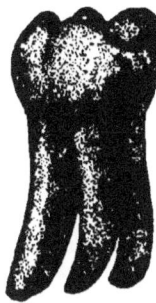
Fig. 82.

FIG. 78 * (Par. 105).—LEFT LOWER FIRST MOLAR, LINGUAL SURFACE. *a*, Mesio-buccal cusp; *b*, disto-buccal cusp; *c*, distal cusp; *d*, mesio-lingual cusp; *e*, disto-lingual cusp; *f*, *f*, gingival line; *g*, mesial root; *h*, distal root. The roots are spread wide apart; *i*, lingual groove.

FIG. 79 * (Par. 105).—LEFT LOWER FIRST MOLAR, LINGUAL SURFACE. This tooth has prominent cusps, and the roots are straight and close together. References the same as for Fig. 78.

FIG. 80 * (Par. 106).—LEFT LOWER FIRST MOLAR, MESIAL SURFACE. *a*, Mesio-buccal cusp; *d*, mesio-lingual cusp; *e*, mesial surface, point of concavity; *f*, *f*, gingival line; *g*, mesial root with broad groove.

FIG. 81 * (Par. 107).—LEFT LOWER FIRST MOLAR, DISTAL SURFACE. The cusps are very prominent. *a*, Mesio-buccal cusp; *b*, disto-buccal cusp; *c*, distal cusp; *d*, mesio-lingual cusp; *e*, disto-lingual cusp; *f*, gingival line; *g*, mesial root; *h*, distal root.

FIG. 82 * (Par. 108).—LOWER FIRST MOLAR, with three roots.

* Illustration, 1½ diameters.

The gingival curvature is generally marked, and the buccal and lingual lines convex. The buccal line is more convex than the lingual, and its curvature is at such an incline as to render the gingival line much longer than the occluding margin. The surface is slightly convex, though almost flat; but, in the central portion, near the gingival line, it is sometimes slightly concave from buccal to lingual. It is rounded away toward the labial and lingual angles. With the occluding surface it forms a sharp angle in the central portion, but is rounded toward the buccal and lingual angles. In the direction of the long axis of the tooth, there is usually a concavity at the junction of the enamel and cementum.

107. The distal surface (Fig. 81) is smoothly convex from buccal to lingual. From the occluding surface to the gingival line it is straight or slightly convex, but forms a considerable concavity at the junction of the crown with the root, which occasionally forms a sharp angle, but generally is well rounded. The occluding margin is often deeply notched by the distal groove. In a few examples the disto-buccal groove is deep after passing over the marginal ridge, and in its distal inclination forms a slight concavity at the disto-buccal angle.

108. The root of the lower first molar is divided into two prongs (Figs. 76 to 82), and this division is usually close to the crown, closer than in any other tooth in the mouth. The mesial root (*g*) inclines first to the mesial, and then curves regularly toward the distal. It is broad from buccal to lingual, and is much thinned from mesial to distal, so that in cross sections it measures double as much one way as the other. It is usually slightly concave on both mesial and distal surfaces, and tapers regularly, but not rapidly, from the bifurcation to the apex, and ends in a flattened, but well rounded, point. The distal root (*h*) inclines to the distal at first, and afterward is nearly or quite straight. In some examples the apical half, curves to the distal, but more often

toward the mesial, so that the apexes of the two roots are inclined toward each other. It is narrower from buccal to lingual than the mesial root, and more nearly round, being but rarely concave or grooved on either mesial or distal surface. It tapers quite regularly, and more nearly to a point than the mesial root, though the apex is generally well rounded. The form of the root is regular, rarely deviating much from its type. Occasionally, however, the division of the root is incomplete. The mesial root is occasionally divided, giving the tooth three roots (Fig. 82); and I have seen a few in which the distal root was divided also, giving four roots.

THE LOWER SECOND MOLAR.

109. The most characteristic difference between the lower first and second molars is the absence of the fifth lobe in the second, and the general change of form which this absence implies; the other parts of the tooth being similar.

110. When the occluding surface is seen in a line with the long axis of the tooth (Fig. 83), the outline of the crown is nearly a parallelogram, with angles rounded and the lines slightly convex, the buccal most. The summits of the marginal ridges are close on the mesial, distal, and lingual margins, while the buccal is carried over toward the lingual by the inclination of the buccal surface. The central inclines of the marginal ridges form a deep central fossa, in the center of which there is usually a deep pit.

111. There are four developmental grooves, all arising from the central pit. The mesial groove (*l*) runs to the mesial margin and crosses the mesial marginal ridge as a fine line, which is often obliterated by wear. The distal groove (*o*) runs in a similar way to the distal surface. These two grooves divide the occluding surface from mesial to distal, and centrally, between the summits of the buccal and lingual marginal ridges. The buccal groove (*m*) runs from the central pit to the buccal margin, and over it to the buccal

Fig. 83.

Fig. 85.

Fig. 86.

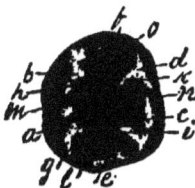

Fig. 84.

Fig. 87.

Fig. 88.

Fig. 89.

FIG. 83 * (Par. 110).—RIGHT LOWER SECOND MOLAR, OCCLUDING SURFACE, of typical form. *a*, Mesio-buccal cusp; *b*, disto-buccal cusp; *c*, mesio-lingual cusp; *d*, disto-lingual cusp; *e*, mesial marginal ridge; *f*, distal marginal ridge; *g*, mesio-buccal triangular ridge; *h*, disto-buccal triangular ridge; *i*, mesio-lingual triangular ridge; *k*, disto-lingual triangular ridge; *l*, mesial groove; *m*, buccal groove; *n*, lingual groove; *o*, distal groove.

FIG. 84 * (Par. 113).—RIGHT LOWER SECOND MOLAR, OCCLUDING SURFACE, of imperfect form. *a*, Mesio-buccal cusp; *b*, disto-buccal cusp; *c*, mesio-lingual cusp; *d*, disto-lingual cusp; *e*, mesial marginal ridge; *f*, distal marginal ridge; *g*, mesio-buccal triangular ridge; *h*, disto-buccal triangular ridge; *i*, mesio-lingual triangular ridge; *k*, disto-lingual triangular ridge; *l*, mesial groove; *n*, lingual groove; *o*, distal groove.

FIG. 85 * (Par. 115).—LEFT LOWER SECOND MOLAR, BUCCAL SURFACE. *a*, Mesio-buccal cusp; *b*, disto-buccal cusp; *c*, buccal groove; *d*, buccal pit; *f*, gingival line; *g*, mesial root; *h*, distal root.

FIG. 86 * (Par. 116).—LEFT LOWER SECOND MOLAR, LINGUAL SURFACE. *a*, Mesio-buccal cusp; *b*, disto-buccal cusp; *c*, mesio-lingual cusp; *d*, disto-lingual cusp; *e*, lingual groove; *f*, gingival line; *g*, mesial root; *h*, distal root.

FIG. 87 * (Par. 117).—LEFT LOWER SECOND MOLAR, MESIAL SURFACE. *a*, Mesio-buccal cusp; *c*, mesio-lingual cusp; *d*, point of proximate contact of mesial surface; *f*, gingival line; *g*, mesial root.

FIG. 88 * (Par. 118).—LEFT LOWER SECOND MOLAR, DISTAL SURFACE. *a*, Mesio-buccal cusp; *b*, disto-buccal cusp; *c*, mesio-lingual cusp; *d*, disto-lingual cusp; *e*, point of proximate contact of the distal surface; *f*, gingival line; *h*, distal root.

FIG. 89 * (Par. 119).—RIGHT LOWER SECOND MOLAR, LINGUAL SURFACE. The roots are curved very much to the distal.

* Illustration, 1½ diameters.

surface, dividing the buccal ridge into two buccal cusps, while the lingual (n) runs to and over the lingual marginal ridge, dividing it also into two lingual cusps. The two divide the tooth from buccal to lingual, into nearly equal parts. The mesial portion is usually slightly the larger. The four grooves form a cross through the occluding surface, dividing it into four lobes, or developmental parts, on each of which there is a cusp and a triangular ridge. In some examples the grooves do not exactly meet at the central pit. The central may rise to the mesial of the buccal, or *vice versa;* or, a similar variation may occur in the central ends of the mesial and distal, causing irregularity of the contour of the central fossa. Occasionally, the lobes are of unequal size, or the cusps are unequally developed, giving rise to imperfect forms (Fig. 84).

112. On the average, the cusps of the second lower molar are higher and more pointed, and the triangular ridges are more prominent than in the first molar. The mesio-buccal and mesio-lingual cusps are generally a little larger than the disto-buccal and disto-lingual; also, the mesio-buccal and mesio-lingual triangular ridges (Fig. 83, *g, i*) are usually more prominent. The crests of these ridges do not run directly toward the central pit, but the two mesial ones meet mesially of the central pit, while the two distal meet distally of it. When they are high, they form transverse ridges by their junction, which separate a mesial and a distal supplemental fossa from the central fossa. When this occurs, there are usually triangular supplemental grooves deflected from the mesial and distal grooves, to the mesial and distal of the triangular ridges, which run towards the angles of the tooth and separate the triangular from the marginal ridges, widening and deepening the supplemental fossæ. A deep pit is often found at the point where these triangular grooves arise from the principal grooves. Supplemental fossæ occur much more frequently, or are more pronounced, in the mesial

than in the distal portion of the tooth, but in many examples the triangular ridges are so widely divided by sulcate, mesial and distal grooves that no supplemental fossæ are seen.

113. In some examples the lower second molar presents differences in the comparative size of its lobes, and the grooves may be deflected from their normal course. Occasionally, the distal groove is divided, and passes over the distal marginal ridge in two divisions, with a small tubercle between them. In poorly developed teeth there may be many supplemental grooves, or wrinkles, running from the developmental grooves up onto the central inclines of the ridges and cusps.

114. Fissures occur oftenest near the central ends of the grooves; though they may appear in any part of their length; and in poorly developed teeth the supplemental grooves may be deeply fissured.

115. The buccal surface of the lower second molar (Fig. 85) is convex in all directions, except that it is partially divided into two sections, or ridges, by the buccal groove (c), which runs over onto it from the occluding surface. In many, this groove ends near the center of the surface in a deep buccal pit (d). This tooth has no disto-buccal groove. The mesial and distal margins converge less toward the gingival line than it does in the lower first molar. The gingival line is nearly straight, and there is a strong inclination of the border of the enamel toward it, giving the appearance of a gingival enamel ridge.

116. The lingual surface (Fig. 86) is similar in all points to that of the lower first molar (q. v., 105); but, on account of a much less convergence of the mesial and distal surfaces toward the lingual, this surface is nearly as great as the buccal surface.

117. The mesial surface (Fig. 87) of the lower second molar is generally a little more convex than in the lower first (q. v., 106); but in other respects they are similar.

118. The distal surface (Fig. 88) differs from that of the lower first molar in not having the distal protuberance of the fifth cusp. Its proximating point with the tooth distal to it is usually central, or toward the lingual, instead of the buccal angle, as in the lower first molar. This surface is usually quite regularly, and smoothly, convex, and its gingival line seldom shows any bucco-lingual curvature.

119. The roots of the lower second molar (Figs. 85 to 89) are similar to those of the first molar; but the divisions are much less spread and less grooved on the mesial and distal sides. In many examples there is but a single root, which is deeply grooved on its buccal and distal sides, marking out the divisions. The root is much more irregular in form than in the lower first molar, and is often much curved distally (Fig. 89) or otherwise distorted.

THE LOWER THIRD MOLAR.

120. The lower third molar, called also the wisdom tooth and *dens sapientia*, is the eighth from the median line, and the last tooth in the arch. It proximates the lower second molar by its mesial surface. The tooth has two typical forms; the one is a four-cusped tooth, similar to the lower second molar (Fig. 83); the other a five-cusped tooth, similar to the lower first molar (Fig. 75); but there are great variations from both of these. Indeed, within the observation of the author the form of this tooth is oftener distorted than any other.

121. The four-lobed tooth is the more common form, and, when well developed, the occluding surface is similar to that of the lower second molar. A supplemental fossa is often seen, formed by the prominence of the mesio-buccal and mesio-lingual triangular ridges; but in the distal portion of the crown such a fossa rarely appears. Indeed, in the four-cusped lower third molars, the distal lobes are generally much smaller than the mesial lobes.

122.   The course of the grooves is often much distorted, so that their central ends fail to proximate, as in Fig. 90, or otherwise.   This renders the form of the central fossa extremely irregular.   Or the principal grooves may be so confused among a number of supplemental grooves that the real dividing lines of the lobes can scarcely be made out (Fig. 91).   In many of these, some of the supplemental grooves run over the marginal ridges, corrugating them, or dividing them into several imperfect cusps.   Occasionally the marginal ridges are nearly equal in height all around the margin of the central fossa, and the enamel surface of the latter is covered with small wrinkles, some of which may be deeply fissured.

123.   In some instances the lower third molar is very large, and in these the ridges may be subdivided into six, seven, or eight cusps, and as many fairly distinct lobes; or, one or more supplemental ridges may appear within the limits of the central fossa surrounded by grooves, which divide them from other parts of the crown (Fig. 92).   Such teeth are usually poorly formed, and the grooves deeply fissured.

124.   The five-lobed lower third molars are very large teeth, larger than the second molars, and very regularly formed.   The distal lobe is placed further to the distal and lingual, and the buccal surface is more rounded than in the first molars.   This form is bilateral and hereditary.

125.   The buccal surface of the lower third molar (Figs. 93, 94, 95) is usually more convex than in the other lower molars, but otherwise of the same form.   If four-lobed, this tooth has the same grooves and pits as the lower second; if five-lobed, it has the markings of the lower first molars.

126.   The mesial, lingual, and distal surfaces correspond with those of the other lower molars, only rather more rounded; especially the distal, which is often nearly a true circle from buccal to lingual.

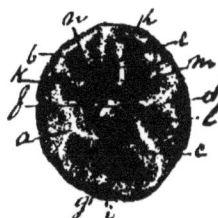

Fig. 90.　　　　　　　　　Fig. 91.　　　　　　　　　Fig. 92.

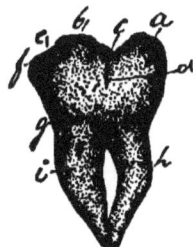

Fig. 93.　　　　　　　　　Fig. 94.　　　　　　　　　Fig. 95.

FIG. 90 * (Par. 122).—LEFT LOWER THIRD MOLAR, OCCLUDING SURFACE, imperfect form. *a*, Mesio-buccal cusp; *b*, disto-buccal cusp; *c*, mesio-lingual cusp; *d*, disto-lingual cusp; *e*, mesial marginal ridge; *f*, distal marginal ridge; *g*, mesial groove; *h*, buccal groove; *i*, lingual groove; *k*, distal groove. The buccal and lingual grooves do not meet in the central fossa as in regular forms.

FIG. 91 * (Par. 122).—LOWER THIRD MOLAR, OCCLUDING SURFACE, very imperfect form. *a*, Mesio-buccal cusp; *b*, disto-buccal cusp; *c*, mesio-lingual cusp; *d*, disto-lingual cusp; *e*, mesial marginal ridge; *f*, distal marginal ridge; *g*, mesial groove; *h*, buccal groove; *i*, lingual groove; *k*, distal groove. Several of the grooves are fissured and are irregular in form.

FIG. 92 * (Par. 123).—RIGHT LOWER THIRD MOLAR, OCCLUDING SURFACE, very large and irregular form. *a*, Mesio-buccal cusp; *b*, a very imperfect disto-buccal cusp; *c*, mesio-lingual cusp, with a sharp, triangular ridge, running very much to the distal; *d*, disto-lingual cusp, standing very much to the mesial of its proper position; *e*, *f*, a large supplemental ridge, occupying the middle portion of the central fossa; *g*, mesial marginal ridge; *h*, distal marginal ridge; *i*, mesial groove, deeply fissured and with supplemental grooves, also fissured, extending to the labial and lingual, forming a supplemental mesial fossa; *k*, buccal groove; *l*, lingual groove; *m*, *n*, deep fissure on either side of the supplemental ridge. There are a number of deep wrinkles running over the distal marginal ridge.

FIG. 93 * (Par. 125).—RIGHT LOWER THIRD MOLAR, BUCCAL SURFACE, with five cusps. *a*, Mesio-buccal cusp; *b*, disto-buccal cusp; *c*, buccal groove; *d*, buccal pit; *e*, disto-buccal groove; *f*, distal cusp; *g*, gingival line; *h*, mesial root; *i*, distal root. In this tooth the lingual root is the larger, and the two come together at their apexes, both of which is unusual.

FIG. 94 * (Par. 127).—RIGHT LOWER THIRD MOLAR, with the roots curved very much to the distal.

FIG. 95 * (Par. 127).—LEFT LOWER THIRD MOLAR, BUCCAL SURFACE; three roots. *a*, Mesio-buccal cusp; *b*, disto-buccal cusp; *c*, distal cusp; *d*, buccal groove; *f*, gingival line; *g*, mesial root; *h*, distal root; *i*, supernumerary root.

* Illustration, 1½ diameters.

127. The root of the lower third molar (Figs. 93, 94, 95), as compared with its crown, is usually much smaller than in the other lower molars. It may be single, or divided into two or more prongs, the tendency being to the formation of two roots, the same as in the other lower molars; and much the greater number has the root in this form, though the single root is common, and three roots (the mesial being again subdivided) are not rare. The root, or roots, of this tooth usually curve distally, sometimes very much, and are otherwise distorted. In extracting, this incline of the roots should be kept in mind.

THE DECIDUOUS TEETH.

128. These are the teeth of early childhood, and serve for mastication till the maxillary bones are sufficiently developed to accommodate the permanent, the larger teeth of adult age. They are then removed by absorption of their roots, which allows their crowns to fall away. Hence they are often called temporary teeth. The shedding process begins about the seventh year, and is completed at from the twelfth to the fourteenth, the succedaneous teeth taking the places of the deciduous. There are twenty deciduous teeth; ten in each jaw, namely: Two central incisors, two lateral incisors, two cuspids, and four molars. This may be expressed by the following formula:

$$I \tfrac{2}{2} \; C \tfrac{1}{1} \; M. \tfrac{2}{2} = 20.$$

There are no bicuspids in the deciduous set, and therefore the deciduous first molars proximate directly with the deciduous cuspids. The bicuspids of the permanent set are succedaneous to the deciduous molars.

129. The incisors and cuspids of the deciduous set are (Figs. 96 to 101) similar in form and lobal construction with their succedaneous teeth, but the deciduous molars give place to the permanent bicuspids, which are of very dissimilar

pattern. The deciduous second molars (Fig. 102, 103), both upper and lower are of the same form and lobal construction as the permanent first molars. The deciduous first molars, upper and lower, have no similar teeth in the permanent set. Their form, and the arrangement of their lobes are peculiar to themselves. Therefore the crowns of these will be separately described.

130. Though the crowns of the temporary teeth, with the exceptions named, are of similar form and lobal construction as the permanent, there are certain minor differences which distinguish them. They are considerably smaller than the corresponding permanent teeth. This reduction in size includes the whole tooth, and is such that it leaves the general proportions unchanged, except that the roots are proportionally longer.

131. The deciduous teeth are, however, marked with a much greater constriction at their necks. The enamel, instead of thinning away to the gingival border as in the permanent teeth, retains its thickness almost to the gingival line and terminates abruptly, leaving a sudden constriction of the neck of the tooth. This varies in degree, but is common to all of the deciduous teeth, and distinguishes them from the permanent teeth.

132. The buccal and lingual surfaces of the deciduous molars slope inward toward the occluding surface much more than those of the permanent, so that the immediate occluding surface, is narrower in proportion to the greatest labio-lingual thickness of the crown. This gives the crown as seen in the mouth the appearance of being very long from mesial to distal; though this characteristic is less marked in the upper than in the lower deciduous molars (see Figs. 104, and 106).

133. The enamel of the deciduous teeth is usually whiter than the permanent teeth, and they are probably of coarser texture. The difference in color is often strongly con-

Fig. 96. Fig. 97. Fig. 98. Fig. 105. Fig. 102.

Fig. 99. Fig. 100. Fig. 101. Fig. 107. Fig. 103.

 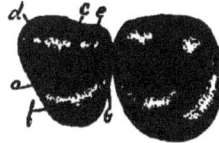

Fig. 106. Fig. 104.

FIG. 96 * (Par. 129).—DECIDUOUS UPPER CENTRAL INCISOR.
FIG. 97 * (Par. 129).—DECIDUOUS UPPER LATERAL INCISOR.
FIG. 98 * (Par. 129).—DECIDUOUS UPPER CUSPID.
FIG. 99 * (Par. 129).—DECIDUOUS LOWER CENTRAL INCISOR.
FIG. 100 * (Par. 129).—DECIDUOUS LOWER LATERAL INCISOR.
FIG. 101 * (Par. 129).—DECIDUOUS LOWER CUSPID.
FIG. 102 * (Par. 129).—LEFT UPPER SECOND DECIDUOUS MOLAR.
FIG. 103 * (Par. 129).—LEFT LOWER SECOND DECIDUOUS MOLAR, BUCCAL SURFACE.

FIG. 104 * (Par. 131).—LEFT UPPER FIRST AND SECOND DECIDUOUS MOLARS. The second deciduous molar has its grooves and lobes in the same form as those of the first permanent molar, Fig. 51. The upper first deciduous molar has but three cusps. a, Mesial groove; b, distal groove; c, buccal groove; d, mesio-buccal cusp; e, disto-buccal cusp; f, lingual cusp.

FIG. 105 * (Par. 138).—LEFT UPPER FIRST DECIDUOUS MOLAR, BUCCAL SURFACE. a, Bucco-gingival ridge; b, mesio-buccal cusp; c, disto-buccal cusp; d, buccal ridge; e, buccal groove.

FIG. 106 * (Par. 141).—LOWER FIRST AND SECOND DECIDUOUS MOLARS, OCCLUDING SURFACES. The lobes and grooves of the lower second deciduous molar are the same as those of the permanent lower first molar, Fig. 75. The lower deciduous first molar has four lobes. a, Mesial groove; b, buccal groove; c, lingual groove; d, distal groove; e, mesio-buccal cusp; f, disto-buccal cusp; g, mesio-lingual cusp; h, disto-lingual cusp; i, mesial fossa.

FIG. 107 * (Par. 118).—LEFT LOWER FIRST DECIDUOUS MOLAR, BUCCAL SURFACE. a, Bucco-gingival ridge; b, mesio-buccal cusp; c, disto-buccal cusp; d, buccal ridge; e, buccal groove.

* Illustration, 1½ diameters.

G

trusted when some of the permanent teeth, as the central incisors, have taken their places by the side of the remaining temporary teeth.

UPPER FIRST DECIDUOUS MOLAR.

134. The occluding surface of the upper first deciduous molar (Fig. 104) when seen in the line of the long axis of the tooth, presents an irregular quadrangular form in which the buccal line is the longer. The mesio-buccal angle is acute, the mesio-lingual is obtuse, and both distal angles are nearly right angles. The buccal margin is irregularly convex, and the lingual margin regularly rounded. Both the buccal and the lingual surfaces are much inclined centrally; or toward the occluding surface.

135. This tooth has three lobes, divided by three grooves. The mesial (a) and distal (b) grooves, run from the mesial to the distal margin in a deep sulcus and divide the lingual from the buccal lobes. Their junction is in a pit in the central fossa. The buccal groove (c) rises from the same pit, at the junction of the mesial and distal grooves, and runs over the buccal marginal ridge to the buccal surface. This groove is generally without a sulcus; or, at most, there is but a slight furrow.

136. The buccal marginal ridge is a high cutting edge which rounds up from the mesio-buccal angle and runs to the distal and buccal till it reaches the point of the mesio-buccal cusp (d). Its direction is then to the distal, descending slightly to the buccal groove, then horizontally to the disto-buccal angle of the tooth to join the distal marginal ridge, the latter portion forming a small disto-buccal cusp (e). In unworn teeth, the buccal groove causes a marked but slight depression where it crosses the ridge, breaking it into two cusps; the mesial being the larger and more pointed. This division is generally defaced very early by wear, so that the ridge presents an almost straight rounded edge.

137. The lingual cusp (f) is in the form of an elevated crescentic edge with its convexity to the lingual, which runs from the mesial termination of the mesial groove (a) to the distal terminatin of the distal groove (b). The central and the lingual inclines of this cusp are nearly equal slopes, while the buccal incline to the buccal marginal ridge is less abrupt than the central. The mesial and distal marginal ridges are not marked by more than a very slight thickening of the enamel, and are cut through by the mesial and distal grooves.

138. The buccal surface (Fig. 105) is remarkable for its bucco-gingival ridge (a), which stands boldly out from the gingival line from one to three millimeters and extends from the mesial to the distal angle. At the mesio-buccal angle it terminates abruptly in a marked prominence, and diminishes gradually as it passes from the mesial to the disto-buccal angle.

139. From the summit of the bucco-gingival ridge to the summit of the buccal marginal ridge, or the mesio-buccal cusp, is nearly a flat surface, except a slight depression along the buccal groove. In many examples there is a slight concavity extending from mesial to distal along the occlusive margin of the gingival ridge, and from the point of the mesio-buccal cusp a strong ridge runs to the mesial prominence of the gingival ridge.

140. The mesial and distal surfaces are quite smoothly flattened. The lingual surface is convex. The neck presents the characteristic constriction common to deciduous teeth.

LOWER FIRST DECIDUOUS MOLAR.

141. The occluding surface of the lower first deciduous molar (Fig. 106), when viewed in a line with the long axis of the tooth, presents the outline of a parallelogram, modified by the rounding of its angles and more or less convexity of its lines. In many, the distal portion of the tooth is broader than the mesial, giving the tooth an ovoid outline. There

are two fossæ. The principal fossa occupies nearly three-fourths of the distal portion of the surface, while the small mesial fossa occupies the immediate mesial portion.

142. The tooth has four lobes of irregular form, divided by four grooves. These grooves all run from the principal fossa. The mesial groove (a, a) rises from the central pit and runs to the mesial, passing over the transverse ridge into the mesial fossa, where it is deflected sharply to the lingual, passing over the marginal ridge near the mesio-lingual angle. This groove varies considerably in its course in different examples. In the principal fossa it usually inclines to the buccal and then toward the lingual, but there is generally an angle at the origin of the buccal groove. The buccal groove (b) rises from the mesial groove, some distance to the mesial of the pit, and runs over the buccal ridge onto the buccal surface, in a slight sulcus, dividing the buccal marginal ridge into two cusps, the mesio- and disto-buccal. Its position determines the relative size of the buccal lobes. The lingual groove (c) runs from the central pit over the lingual marginal ridge onto the lingual surface, and is deeply sulcate on the central incline of the ridge.

143. This tooth has four cusps corresponding with the four lobes. The mesio-buccal lobe (e) is very irregular in its outline. It forms the entire mesial marginal ridge, and from one-third to three-fourths of the slopes of the mesial fossa. The mesial marginal ridge is usually high in young, unworn, teeth. It is curved, and from the mesio-buccal angle it becomes the buccal marginal ridge, and rises to the distal to form the point of the mesio-buccal cusp (e). From the point of the cusp it falls away to the distal and buccal to the buccal groove. A prominent triangular ridge descends from this cusp to the lingual and distal, and joins a similar triangular ridge from the mesio-lingual cusp (g) to form the transverse ridge, thus dividing the mesial from the principal fossa. Exceptionally, a deep sulcus divides these triangular

ridges and connects the fossæ. From the buccal groove, the
buccal marginal ridge passes almost directly to the disto-
buccal angle where it joins the distal marginal ridge. In the
central portion it rises slightly to form the low disto-buccal
cusp (*f*). The triangular ridge from this cusp is usually low,
or wanting.

144. The lingual marginal ridge (*g, h*) rises abruptly
from the mesial groove to the summit of the mesio-lingual
cusp (*g*), and then falls away from the distal and lingual to
the lingual groove. In most specimens, in unworn teeth,
the mesio-lingual cusp is sharp, and its point is carried by
the lingual incline far toward the central line of the tooth,
so much so as to be in marked contrast with the general
form of the lingual cusps of the molars. From its apex a
triangular ridge descends to join that from the mesio-buccal
cusp in forming the transverse ridge. From the lingual groove,
the lingual marginal ridge rises to the point of the disto-
lingual cusp (*h*) and then falls away in a curve to form the
distal marginal ridge. This cusp is generally rather low, but
varies much in these teeth. In some cases, there is a sharp
triangular ridge descending into the central fossa, but more
generally this ridge is slight.

145. The distal marginal ridge is usually made to appear prominent by the depth of the principal fossa. It is
crossed near its center by the distal groove.

146. The principal fossa is generally deep and well
rounded. The distal triangular ridges, the only ones descending into this fossa, are generally not prominent, but occasionally they are sufficiently so to render the fossa very angular.
I have observed many in which the enamel in this fossa was
very imperfect and the bottom of the fossa broad and rough.

147. The mesial fossa (*i*) is usually sharp and deep with
smooth sides and a central pit that is frequently the seat of
caries.

148. The buccal surface of the lower first deciduous

molar (Fig. 107) is remarkable for its prominent bucco-gin-gival ridge (a), which runs from the mesio-buccal to the disto-buccal angle, and stands out prominently over the junction of the crown with the root. From mesial to distal, this ridge slopes toward the occluding surface, making the crown longer at the mesial angle than at the distal. From this ridge the surface slopes rapidly toward the occluding margin, and more rapidly at the mesial than the distal portion. From the mesio-buccal cusp a strong ridge of enamel runs to the more prominent portion of the gingival ridge near the mesio-buccal angle. Otherwise this surface is nearly flat. The lingual surface is usually well rounded, but is broken toward the occluding surface by the prominence of the mesio-lingual cusp, and the sulcus of the lingual groove.

149. The mesial and distal surfaces are slightly rounded. The disto-buccal and lingual angles are about equal in prominence, but the mesio-buccal and lingual angles are very unequal. The mesial surface slopes rapidly to the lingual, making the lingual surface much shorter than the buccal. The mesio-buccal angle is acute and prominent, while the mesio-lingual is very obtuse and rounded.

150. The root of this molar is divided into two prongs which are spread widely apart. They are thin from mesial to distal, and slightly grooved; and, from buccal to lingual, broad. They taper regularly to broad flat apexes, which are occasionally bifurcated near their apexes.

151. The roots of the deciduous teeth are the same in number and general characteristics as in the teeth of the same denomination in the permanent set, except that those of the molars are more divergent. This spreading of the roots accommodates the crowns of the permanent bicuspids, which are developed between the roots of the deciduous molars. Those of the lower jaw are thin from mesial to distal, broad from labial to lingual, and grooved along their flattened sides. The mesial and distal roots of the upper deciduous molars.

are, also, thin, grooved, and widely divergent. The lingual root stands boldly to the lingual, forming a wide space between the three, for the crowns of the upper bicuspids. In many examples the lingual and distal roots are joined by broad thin connections for the greater part of their length.

THE PULP CHAMBER.

152. Every tooth has a cavity in the center of the crown, and one or more canals extending through the long axis of the root, or roots, to the apex. This cavity contains a tissue composed of cellular elements imbedded in a semi-gelatinous matrix, filling every part of the space, and is richly supplied with blood-vessels and nerves. This is known as the pulp of the tooth.*

153. The central cavity in the tooth is usually divided into a crown, or coronal portion, and a root portion. Its parts are familiarly known as the pulp chamber (crown cavity), and root canal, or root canals. The pulp chamber is comparatively large, and the root canals are small, tapering from the pulp chamber to a minute opening at the apex of the root, known as the apical foramen. In those teeth that have prominent cusps, as the bicuspids and molars, there is a prolongation of the pulp toward the point of each cusp. These are known as the horns of the pulp; and the prolongations of the chamber are designated the horns of the pulp chamber.

154. The size of the pulp chamber and of the root canals varies greatly in teeth of different denominations; and also in different teeth of the same denomination. In the early formative stages of the teeth, it is very large, and diminishes in size as growth proceeds, until the tooth is fully formed. Afterward this diminution goes on slowly,

---

*As the form of the pulp chamber gives the exact form of the pulp, no separate description of the pulp will be given. It is not my purpose to give any histological descriptions in this work.

until, in old age, it is often nearly obliterated.. In the formative stage, *i. e.*, during the growth of the root of the tooth, the root canal is large and funnel-shaped, with the open end of the funnel toward the apex of the root. As growth proceeds, and the root approaches completion, this diminishes rapidly till the root is fully formed, when it is contracted to a small foramen. This, however, continues to diminish slowly. Therefore, the size of the pulp chamber, the root canals, and the apical foramen, are greater in youth than in old age. However, after adult age is reached, the diminution in size is usually not great. During this time, the horns of the pulp chamber are shortened by the same process of formation of dentine on their surface that is going on in all parts of the pulp chamber and root canal. Therefore, the horns of the pulp become shorter, or recede, as age advances. Indeed, the whole pulp, very slowly, becomes smaller.

155. Certain processes, when present, also, serve to diminish the size of the pulp chamber more rapidly; especially abrasion of the teeth, a matter that seems to depend largely upon the character of the occlusion. When the occlusion is such that there is much rubbing, or sliding motion, of the teeth against each other, wear goes on rapidly. This seems to induce depositions of dentine on the walls of the pulp chamber, which reduces its size; and, especially, causes the recession of the horns of the pulp. In this way, exposure of the pulp from the wearing away of the dentine is delayed or prevented. In many instances the pulp chamber is almost obliterated in the molars and bicuspids, and recedes rootwise of the gingival line in the incisors and cuspids. Slowly progressive caries or erosion of the teeth often induce similar deposits.

156. In the incisors and cuspids, the pulp chamber and the root canal are not sharply differentiated. The latter tapers, gradually, from the full size of the largest crown portion to a small foramen at the apex of the root. In teeth

with more than one root, the transition from pulp chamber to root canal is usually sharply defined, the former being very large as compared with the pulpal end of the latter. Indeed, the general form of the pulp is a diminished counterpart of the form of the surface of the tooth, except that it is in every way more slender.

157. Our studies thus far have been of the outer surfaces of the teeth. The pulp chambers are within, and, therefore, in the study of them, dissections must be made to expose them to view. It is often necessary for the dentist to enter the pulp chamber of the teeth of his patients, and there perform delicate operations with a precision which demands the most accurate knowledge of this cavity. Therefore this work of exposure, and examination, of the pulp chambers of teeth should be thorough, as a preparation for operations in the mouth. The form of dissection necessary will be given in connection with the teeth as they are individually described.

THE PULP CHAMBERS OF THE UPPER INCISORS.

158. The pulp chambers and root canals of the upper central, and lateral incisors are so similar, the description of one will do for all.

Dissection.—1st. Saw the tooth through on the gingival line, at the labial surface, at right angles with the long axis.

2d. Saw the tooth from labial to lingual, along the central line of the long axis from end to end.

3d. Saw the tooth from mesial to distal, along the central line of the long axis from end to end.

159. A very thin saw in a strong frame should be used, otherwise the lengthwise dissections should be made to one side of the central line in the first instance, and afterward the whole length of the pulp chamber exposed by grinding on a stone; or the lengthwise exposure may be made by catching the tooth in the vise and removing one-

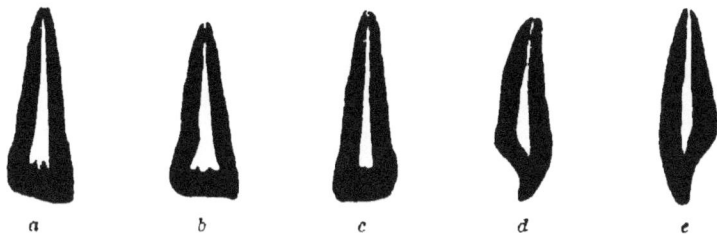

Fig. 108.

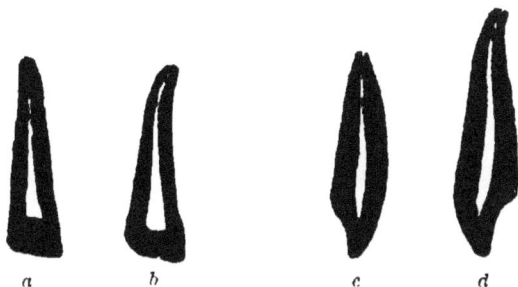

Fig. 109.

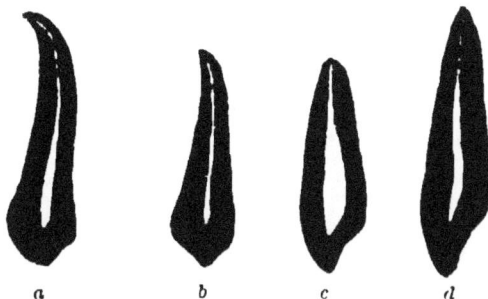

Fig. 110.

FIG. 108 * (Par. 160).—THE PULP CHAMBER OF THE UPPER CENTRAL INCISOR. a, b, Mesio-distal sections of the young teeth, showing the three short horns of the pulp; c, mesio-distal section of a tooth from an adult; d, e, labio-lingual sections.

FIG. 109 * (Par. 160).—PULP CHAMBER OF THE UPPER LATERAL INCISOR.—a, b, Mesio-distal sections; c, labio-lingual sections; d, labio-lingual section of a very long lateral incisor.

FIG. 110 * (Par. 163)—PULP CHAMBER OF THE UPPER CUSPIDS. a, b, Mesio-distal sections; c, d, labio-lingual sections.

* Illustration, actual size.

half with the file, or by grinding on an emery wheel. After
the pulp chamber is exposed so that half of its concavity
remains in the half of the tooth, and has been ground
smooth and flat, it should be inked on an inked pad (such as
is used for rubber stamps for printing), and a print made
from it. This will give the form of the tooth and pulp
chamber in silhouettes similar to the accompanying illustrations. The printing is facilitated by sticking the tooth to
a piece of hard wax for convenience in handling. Ordinary
modeling compound, or sealing wax, is convenient and effec-
tive. The printing is usually better done by laying ordinary
writing paper on a sheet of semi-soft rubber, about one-eighth
inch in thickness. This is specially useful when the ground
surface cannot be perfectly flat, as in curved roots.

160. In the upper central and lateral incisors (Figs. 108
and 109), there is no distinct division of the pulp cavity into
pulp chamber and root canal; but there is one straight canal,
from the interior of the body of the crown to the apex of
the root, of which the crown portion is the larger. In young
teeth, this has very distinctly the form of the surface of the
tooth and root, except that it is much more slender. The
largest diameter of the cavity is about level with the gingival line on the labial surface. From this point, the pulp
chamber, or canal, extends toward the cutting edge of the
tooth, about two-thirds the length of the crown, sometimes
a little more, often less, and ends in a thin edge, broad from
mesial to distal. In young teeth this edge has three short
horns (Fig. 108, a, b), or prolongations, extending toward the
three small cusps seen on the edge of unworn incisors (21).

161. From the gingival line toward the apex of the
root it tapers very gradually and regularly to a narrow canal.
Just within the apex of the root, almost at the end, there is
usually a sudden contraction of the diameter of the canal,
lessening it from one-third to one-half. This is the apical
foramen; but this contraction of the canal is not usually

present for one or two years, or more, after the tooth has taken its place in the arch (q. v., 154).

162. The size of the canal becomes smaller from youth to old age. In incisors just taking their places in the arch, I have found the diameter of the canal at the gingival line to be from one-fourth to one-third the diameter of the neck of the tooth. In early adult age, the canal may be said to average about a fourth the diameter of the neck of the tooth, ranging down as age advances to one-fifth, or sixth, and even to one-tenth. In the lateral incisor, the chamber and canal are a little smaller than in the central, but larger in proportion to the size of the tooth.

THE UPPER CUSPID.

163. The pulp chamber and root canal of the upper cuspid (Fig. 110) is about the same in form as that of the central and lateral incisors, except that the coronal extremity has the central horn much extended toward the apex of the cusp of the tooth, and the lateral horns are practically absent. The canal is proportionally somewhat smaller. However, this tooth is often somewhat flattened at the neck, the long diameter being from labial to lingual. In this case the pulp canal at the neck, and from thence toward the apex of the root, is also much flattened in the same direction, but is progressively rounded as the apex is approached. In some examples, the labio-lingual diameter of the canal is double the mesio-distal. As age advances, and the canal becomes smaller, the opening is occasionally reduced to a mere slit.

PULP CHAMBER OF THE LOWER INCISORS.

164. (Fig. 111). The coronal portion of the pulp chamber of the lower incisors is much flattened. At the level of the gingival line, the long diameter is from labial to lingual. The chamber extends toward the cutting edge of the tooth, about two-thirds the length of the crown, and in

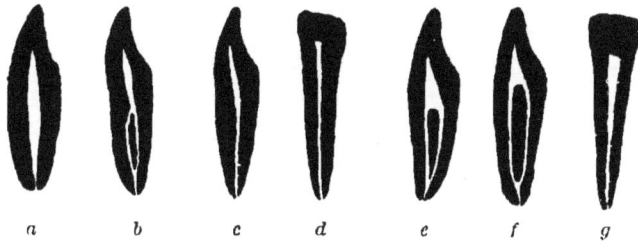

Fig. 111.

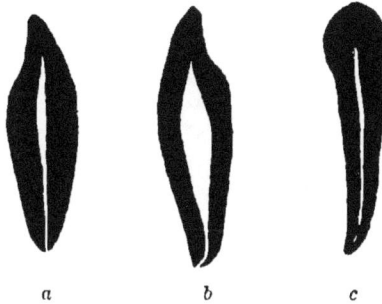

Fig. 112.

FIG. 111* (Par. 164).—PULP CHAMBER OF THE LOWER CENTRAL AND LATERAL INCISORS. a, b, c, Labio-lingual sections, showing differences of form of the pulp chamber ; d, mesio-distal section ; e, f, labio-lingual sections, showing the more usual forms of the pulp chamber ; g, mesio-distal section, showing large pulp chamber.

FIG. 112* (Par. 165).—PULP CHAMBER OF THE LOWER CUSPIDS. a, Labio-lingual section, showing a small pulp chamber; b, labio-lingual section, showing a very large pulp chamber; c, mesio-distal section.

* Illustration, actual size.

this extension its diameter is progressively diminished from labial to lingual, and extended from mesial to distal, following the contour of the surface of the tooth, and ends in a thin edge. In young teeth this has three short projections toward the slight cusps seen on the cutting edges of the young, unworn teeth. The root has usually a narrow slit-like opening for the greater portion of its length (d, g), corresponding with the form of the flattened roots. In many instances, however, the root canal is divided into two portions, or canals, for a part of its length (e, f). In the adult, these canals are usually very small, the point of separation into two canals being irregular, but usually slightly root-wise from the level of the gingival line. It may occur at about the level of the gingival line, or the canal may remain single for half the length of the root, and then be divided for a space, the two uniting again before reaching the apex. Generally, there is but one apical foramen. Instances occur in which there are two, each canal remaining distinct to the end. As age advances, the canals of the lower incisors often become very minute.

PULP CHAMBER OF THE LOWER CUSPID.

165. The pulp chamber, and the root canal of the lower cuspid (Fig. 112) are variable in size and form. At the neck of the tooth the chamber is usually irregularly flattened, with the longer diameter from labial to lingual, and the labial portion wider than the lingual. The coronal portion extends about two-thirds of the length of the crown toward the point of the cusp, ending in a point, or horn, which is often very slender. The form of the root portion of the canal depends on the form of the root. It is sometimes nearly round, but more frequently it is sharply flattened for the greater portion of its length, becoming more rounded toward the apex. Occasionally, this canal is divided for a part of the length of the root. Also, the root is sometimes

H

divided, a very small prong appearing on its lingual side. In this there is usually a very small canal that is difficult to enter with a broach. In some lower cuspids the canal is very small (*a*), in others, very large (*b*). In a few instances I have seen it more than one-third the diameter of the root in the adult. In this case the diameter of the canal is greater than the thickness of the walls of dentine and cementum by which it is inclosed. This renders the pulp very liable to exposure in excavating carious cavities.

PULP CHAMBERS OF THE BICUSPIDS.

166. *Dissections.*—1st. Divide the crown from the root on the gingival line with a fine saw.

2d. Divide the tooth from labial to lingual through its length with a fine saw; or remove the distal half of the tooth with the file, or stone. These two dissections will usually exhibit the pulp chamber and root canals sufficiently, though in the single-rooted teeth with two canals it is better to divide the root crosswise at the middle of its length, or at several points.

PULP CHAMBER OF THE UPPER FIRST BICUSPID.

167. The pulp chamber and root canals of this tooth differ from those of the incisors and cuspids, by having the coronal chamber distinguished sharply from the root canals (Fig. 113, *d, e*). The chamber is centrally located in the long axis of the crown of the tooth, the lateral walls being about equal in thickness. The center of the pulp chamber is about level with the gingival line, or a little toward the occluding surface. The occluding walls are thicker than the lateral, and vary in thickness from about one-third to two-thirds of the length of the crown of the tooth. The form of the pulp corresponds closely with the form of the tooth. A horn extends from the coronal portion toward the apex of each cusp. The buccal horn rises from the extreme buccal part of

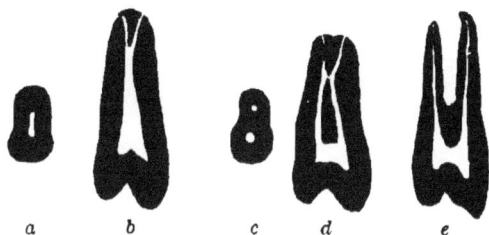

a   b   c   d   e

Fig. 113.

a   b   c   d   e   f

Fig. 114.

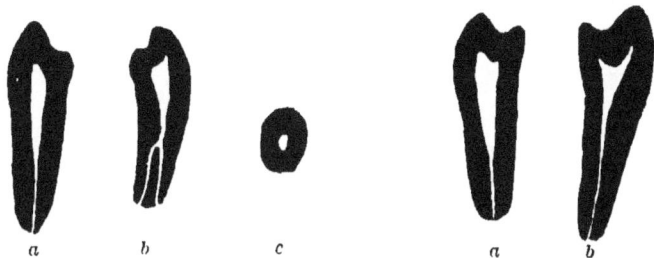

a   b   c   a   b

Fig. 115.   Fig. 116.

FIG. 113 * (Par. 167).—PULP CHAMBER AND ROOT CANALS OF THE UPPER FIRST BICUSPIDS b, Labio-lingual section of single-rooted tooth, with single canal divided near the apex of the root; a, cross-section of the same; d, labio-lingual section of a single-rooted tooth with two canals, which connect at one point; c, cross-section of the same, a little root-wise from the pulp chamber; e, labio-lingual section of a double-rooted tooth, showing the more usual form of the chambers and canals.

FIG. 114* (Par. 168).—PULP CHAMBER AND ROOT CANALS OF THE UPPER SECOND BI-CUSPID. a, Labio-lingual section showing chamber with a long slender horn, also two canals which unite in the apical third of the root; b, cross section of a root with two canals; c, cross section of a root with a single large canal; d, labio-lingual section of a single root with two canals; e, f, labio-lingual sections showing the more usual form of the pulp chamber and root canal of the tooth.

FIG. 115 * (Par. 169).—PULP CHAMBER AND ROOT CANALS OF THE LOWER FIRST BICUSPID. a, Labio-lingual section showing the more usual form; b, labio-lingual section showing a peculiar and very unusual division of the root canal; c, cross section in the body of the root.

FIG. 116 * (Par. 169).—PULP CHAMBER AND ROOT CANAL OF THE LOWER SECOND BI-CUSPID. a, b, Labio-lingual sections showing the more usual forms of the chamber and canal in this tooth.

---

* Illustration, actual size.

the pulp, while the lingual horn rises from the extreme lingual portion. Sometimes, especially in long cusped teeth, they are very long and slender, extending far toward the points of the cusps, and in rare cases almost, or even quite to the enamel. As age advances, they become shorter, and in old age have almost disappeared. In thick-necked teeth with short cusps, the horns of the pulp chamber are short, and the occluding wall is usually very thick.

167. The root canals in bicuspids that have two roots pass from the pulp chamber through the center of each root to its apex, and are known as the buccal and lingual root canals (e). The buccal canal arises from the extreme buccal side of the pulp chamber, and the lingual canal from the extreme lingual side, and their course is almost parallel with the walls of these two portions of the pulp chamber. The floor of the chamber is rounded over in an arch from one canal to the other. Each canal begins in a funnel-shaped opening, which leads into a narrow round canal, which tapers gradually to the apical foramen. Many of the upper first bicuspids have only one root; but they generally have two root canals almost exactly similar to those with two roots. Occasionally, however, these come together and end in one apical foramen, or there may be a communication between the two canals in some part of their course (d). More rarely, the upper first bicuspid has one broad (from buccal to lingual) flat canal passing through the whole length of its single root. Sometimes this is divided near the apex (b).

PULP CHAMBER OF THE UPPER SECOND BICUSPID.

168. The pulp chamber of the upper second bicuspid (Fig. 114) is very similar to that of the first (q. v., 167); but the horns of the pulp are usually shorter. In this tooth there is generally but a single root canal (e, f). This is approached by an opening that is broad from buccal to lingual, and tapers gradually toward the apex of the root, ending in a

narrow apical foramen. The canal is often quite large, and the demarkation of the pulp chamber, as distinguished from the root canal, very indistinct, or entirely absent. Examples are not infrequent, however, in which there are two root canals in the single root (*d*). They are then similar to those of the first bicuspid ; but, sometimes, the two canals end in a common apical foramen (*a*).

PULP CHAMBERS OF THE LOWER BICUSPIDS.

169. The pulp chambers of the lower bicuspids (Figs. 115 and 116) seldom show a marked distinction from the root canals. There is, however, usually a coronal bulbus portion which connects with the pulp canal proper by an extended funnel-shaped constriction (*a, b*). In the lower first bicuspid, the coronal extremity ends in a horn, which extends toward the point of the buccal cusp. This horn may be short and obtuse, or long and pointed. There is, generally, a well marked protrusion toward the lingual cusp, but no extended horn. It is rather an enlargement of the bulb in that direction. In the lower second bicuspid, this protrusion is more considerable, and in some examples it is elongated into a slender point, especially in young teeth (Fig. 116, *b*). In the three cuspids, lower second bicuspids (162, Fig. 49) there are two of these on the lingual side, spreading toward the mesial and distal. They are generally short, but by their protrusion are brought rather nearer the surface of the tooth than other horns of the pulp ; and are, therefore, more liable to be opened into when excavating proximate cavities.

170. The root canals of the lower bicuspids are usually large in the first half, tapering to a fine canal in the apical third, of their length. The canal of the lower first bicuspid is usually nearly round, and the second is considerably flattened ; and in both they are usually straight. Bifurcations of these canals are rare, but occur occasionally. In the illustrations (Fig. 115, *b*) one is shown dividing in such a way

that the division would not be likely to be detected by a broach.

PULP CHAMBERS OF THE UPPER MOLARS.

171. *Dissections.*—1st. With a fine saw, separate the crown from the root level with the gingival line.

2d. Cut away the mesial surface of the crown and the mesial surface of the mesial and lingual roots, with the file or a corundum stone, till the canals in each are fully exposed. As the mesial root is generally curved, some care is required to fully expose the full length of the canal without cutting too far in the central portion of its length. If the curved surface is made smooth, good prints can generally be made by using a piece of semi-soft rubber under the paper, and, while pressing it to the paper, rolling the tooth so as to bring all of the length of the curved surface in contact.

3d. Cut away the buccal surface so as to expose the pulp chamber and canals of the two buccal roots, observing the same precautions as in the second dissection.

In the first dissection, both crown and root should be examined. First, clean the portion of the chamber in the crown to study carefully its horns and its general shape or contour with relation to the outer surface of the tooth. The root canals should be cleaned with the broach, and their size and direction carefully studied; also, the position of the openings leading from the pulp chamber should be studied with regard to their relation to the several points of the surface of the crown. This latter is especially important. Several dissections should be made of each of the upper molars.

4th. Grind away the root portion of the first dissection, printing occasionally till the bifurcation of the roots is reached.

172. The pulp chamber of the upper molars is very distinct from the pulp canals; the latter often leaving the

former by very small openings (Fig. 117). The average diameter of the pulp chamber is about equal to the thickness of the lateral walls by which it is surrounded, sometimes more, sometimes less. The occluding wall is usually considerably thicker. The form of the pulp chamber is generally similar to that of the crown of the tooth; but the horns in the young tooth are often quite slender as compared with the cusps, and penetrate far toward the enamel. The length of these diminish as age advances. In teeth much flattened from mesial to distal, as often occurs in the upper first molars, and especially with the second, the equal thickness of the lateral walls is usually maintained pretty closely, so that the flattening of the pulp chamber seems out of proportion to the form of the tooth.

173.   The floor of the pulp chamber is rounded or arched in the center, and falls away toward the mouths of the canals. The latter are situated in the position of the angles of a triangle (*the molar triangle*) (Figs. 118 and 119), the mesial line of which is the longest, the buccal the shortest, and the distal the intermediate length. For the first molar, this triangle is well shown in the illustrations representing sections a little rootwise from the floor of the pulp chamber (*c*). This is best seen in the specimen itself; and the position and the direction of the canals, with relation to the walls of the pulp chamber and the main points of the surface of the crown, should be carefully studied.

174.   The opening into the lingual root (Fig 117, *d*) is the simplest and most direct. Generally, it begins in a funnel-shaped opening inclining to the lingual, which quickly narrows to the dimensions of a moderately small canal, and continues to taper to the apical foramen. It is usually straight, or but slightly curved.

175.   The opening of the mesial canal is under the mesio-buccal cusp, close against the mesio-buccal angle of the pulp chamber. It often happens that this canal opens in

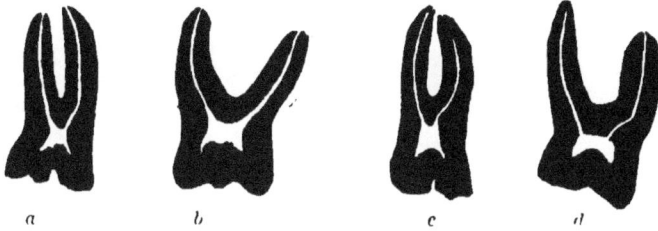

Fig. 117.

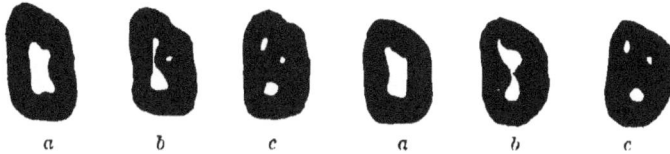

Fig. 118.   Fig. 119.

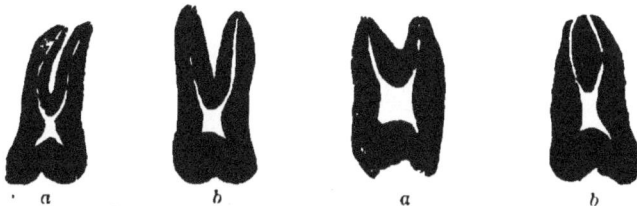

Fig. 120.   Fig. 121.

FIG. 117 * (Par. 172).—PULP CHAMBER AND ROOT CANALS OF THE UPPER FIRST MOLAR. *a*, labio-lingual section showing the chamber and the canals in the distal and lingual roots; *b*, *d*, mesio-distal sections showing the pulp chamber and the canals in the mesial and distal roots; *c*, labio-lingual section showing the pulp chamber and the canals in the mesial and lingual roots.

FIGS. 118,* 119 * (Par. 173).—PULP CHAMBER AND ROOT CANALS OF THE UPPER FIRST MOLARS, cross sections. *a*, Centrally through the pulp chamber; *b*, section just at the floor of the pulp chamber; *c*, section a little rootwise from the pulp chamber, showing the canals, and the form of the molar triangle.

FIG. 120 * (Par. 177).—PULP CHAMBER AND ROOT CANALS OF THE UPPER SECOND MOLAR. *a*, Mesio-distal section showing pulp chamber and the canals in the mesial and distal roots; *b*, labio-lingual section showing the chamber and the canals in the distal and lingual roots.

FIG. 121 * (Par. 178).—PULP CHAMBER AND ROOT CANALS IN THE UPPER THIRD MOLAR. *a*, Labio-lingual section showing pulp chamber and the canals in the mesial and lingual roots; *b*, mesio-distal section of the single rooted tooth showing the form of the pulp chamber and the mesial and distal root canals.

_____

* Illustration, actual size.

a groove in the angle of the chamber (Fig. 119, *b*), making this the thinnest point in the dentinal walls surrounding it. In young teeth, the mouth of the canal is of a flattened funnel-shape, which is quickly contracted into a very fine canal; but in the adult, it often begins as a fine canal. Its course at first is to the buccal and mesial, and then curves to the distal. It is usually distinctly flattened, and often has a thin edge to the lingual. It is often a very difficult canal to clean with a broach. To find this canal the point of the broach should be directed into the mesio-buccal angle of the pulp chamber; and, while held against the wall within this angle, it is slid toward the root. It will rarely fail to glide into the canal.

176. The distal canal usually begins abruptly as a fine opening (Fig. 117, *a*, *c*), situated at the disto-buccal angle of the floor of the pulp chamber (Figs. 118 and 119); so that a broach pressed into that angle will easily glide into it. But in some instances, especially in the upper second molars, the opening is in the floor of the pulp chamber at a little distance from the immediate angle toward the center of the floor, and then, in positions which limit the use of the eye, it is often difficult to find. In teeth much flattened at the neck, the opening of this canal may begin very close to the mouth of the mesial canal (Fig. 120, *a*), or close against the distal wall of the chamber, half way from the buccal to the lingual wall, or, anywhere between this point and the disto-buccal angle. The first direction of the canal will vary according to its position. If it is found in a fairly well-defined disto-buccal angle of the chamber, its direction will be a little inclined to the distal, and the broach will penetrate it easily; if in the floor of the chamber, it will sometimes be straight, as in the former case; but more generally the first direction will be to the distal and buccal, with considerable curve afterward. If found close to the mesial canal, its course is usually first sharply to the distal, when it curves rather

abruptly toward the apex of root. If found along a smooth or curved distal wall, the course will generally be to the distal and buccal, with but little curve. This canal is usually very fine from its beginning, and almost, or quite round.

177. While the canals are similar in all of the upper molars, there are differences in the form of the floor of the pulp chamber that may be briefly generalized. The pulp chamber of the upper second molar (Fig. 120) is usually much more flattened from mesial to distal than in the first molar. This changes the relation of the mouths of the canals somewhat, rendering the distal angle of the triangle formed by them more obtuse, and brings the mouth of the distal canal nearer the mesial line of the triangle, so that it seems to be found along the distal wall of the narrowed chamber. In others, it is found in the extreme buccal portion crowded close against the mouth of the mesial canal.

178. The position of the openings of the canals in the upper third molar (**Fig. 121**) is usually much the same as in the first and second, varying so as to resemble either. Occasionally there is more than the usual number; and others with only one or two. When there is but one, it is commonly quite large. Four, five, or even seven, or eight, are sometimes found.

PULP CHAMBERS OF THE LOWER MOLARS.

179. *Dissections.* 1st. Saw the tooth through the gingival line dividing the crown from the root. This cut will pass through the body of the pulp chamber and give a good view of the roof, and floor, and a good idea of the general form. The root canals should be cleaned and examined with the broach.

2d. Saw the tooth through from end to end centrally from mesial to distal, or grind or file away the buccal side till the pulp chamber and the root canals are exposed. As there are usually two canals in the mesial root, an exact

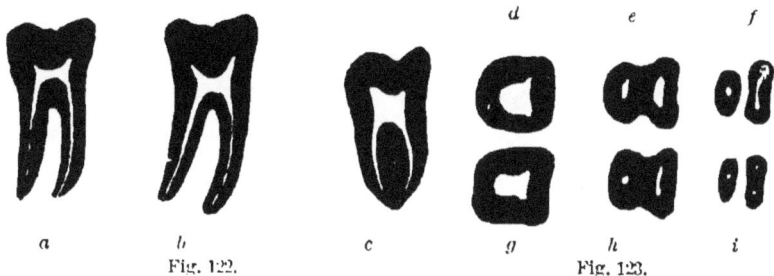

a   b
Fig. 122.

d   e   f

c   g   h   i
Fig. 123.

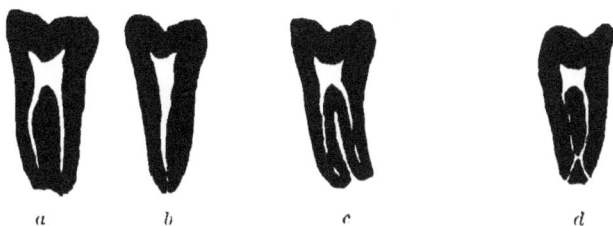

a   b   c   d
Fig. 124.

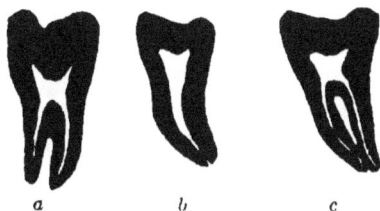

a   b   c
Fig. 125.

FIG. 122* (Par. 180).—PULP CHAMBER AND ROOT CANALS OF THE LOWER FIRST MOLAR. a, b, Mesio-distal sections showing the form of the pulp chamber and root canals; c, bucco-lingual section showing the canals in the mesial root.

FIG. 123* (Par. 180).—CROSS-SECTIONS THROUGH THE CROWN AND ROOT OF THE UPPER FIRST MOLAR, showing the pulp chamber and the root canals. d, g, Sections through the pulp chamber; e, h, sections a little rootwise from the pulp chamber; f, i, sections near apex of root.

FIG. 124* (Par. 180).—PULP CHAMBER AND ROOT CANALS OF THE LOWER SECOND MOLAR. a, c, Mesio-distal sections, showing the form of the pulp chamber and root canals; b, bucco-lingual section of the distal root and crown; d, bucco-lingual section through the mesial root and crown, showing two canals with communication in the apical third of the root. This communication is not very common.

FIG. 125* (Par. 182).—PULP CHAMBER AND ROOT CANALS OF THE UPPER THIRD MOLAR. a, c, In double-rooted teeth; b, single-rooted teeth.

_____

* Illustration, actual size.

central cut from mesial to distal will generally fail to expose the canals, and the cut will be better made at a slight angle, so as to expose either the labial or lingual canal of the mesial root.

3d.   Grind, or file, away the mesial surface of the crown and root till the pulp chamber and the whole length of the canals in the mesial root are exposed. As this root is usually curved, the cutting must be done with care, and the curve followed.

4th.   Cross sections of the roots should be made at intervals. An excellent study is to begin grinding at the apex of the roots, printing occasionally, and continuing the grinding until the pulp chamber is reached. This will display cross sections, at intervals, of the entire root canals. Enough of these dissections should be made, of each of the lower molars, to make the student familiar with each class.

180.   The pulp chamber of the lower molars (Figs. 122, 123, 124) has the same general form as the surface of the crown, but is generally rather more angular. The wall of the chamber toward the occluding surface is convex toward the pulp; the horns extend from the extreme angles toward the apex of each cusp. The floor, through the central portion, is arched or convex from mesial to distal, and concave from buccal to lingual. The mesial wall of the cavity is flat, and longer than the distal. The mesio-buccal and mesio-lingual angles are sharp and projecting, while the distal angles are rounded (Fig. 123, d, g). The size of the chamber varies much. In youth, its diameter is often as much as two-fifths of the crown, and seldom less than one-third. This diminishes as age advances, and in old age, it is often very small, or especially where there has been considerable abrasion of the teeth, the pulp chamber may be almost obliterated.

181.   The root canals of the lower molars proceed from the mesial and distal portions of the pulp chamber (Fig. 122, a, b). The mesial canal, at its mouth, is usually about as

broad from buccal to lingual as the whole breadth of the chamber, including its angular projections. Either at, or a little rootwise from the floor of the pulp chamber, it is usually divided into two very small canals which diverge at first, and approach each other afterward, but usually remain distinct, each ending in its own apical foramen (Fig. 122, c). Occasionally, however, they are united in the apical third of the root, and end in a common apical foramen. Again, there may be a communication between them in the apical portion of the root, each canal remaining otherwise complete in itself. A few, have one broad flattened canal (Fig. 123, d, e, f). These canals are usually minute, and very difficult to thoroughly clean with the broach, though the mesio-buccal canal is usually easily found, if the pulp chamber is thoroughly opened. By placing the point of the broach in the mesio-buccal angle of the chamber and pushing it gently on, it will generally glide into the canal. The first direction inclines to the mesial and buccal, after which it curves to the distal and lingual (Fig. 122, c). Generally, these curves are easy, without short bends. The broach easily glides into the mesio-lingual canal by placing the point in the mesio-lingual angle of the pulp chamber and sliding it toward the root. The first inclination is to the mesial, but occasionally to the lingual, after which it curves to the distal and buccal.

182. The distal canal is approached by a funnel-shaped opening, of which the central part of the distal wall of the pulp chamber becomes a portion. Its direction is a little to the distal, and is generally very nearly straight to the apex. At first, it is flattened, with the long diameter from buccal to lingual, and progressively becomes rounded, and tapers regularly to the apical foramen. It is generally much larger than the canals of the mesial root and is easily cleaned with the broach. If the mouth is wide open, and the handle of the broach brought against the upper central incisors with

the point directed against the posterior wall of the pulp chamber, it will easily glide into the canal, and pass to the apical foramen. This particular position for easily entering the distal canal is important, for all the lower molars. Occasionally, the lower third molar has but one root canal (Fig. 125, *b*) and is then generally very large. More rarely, only a single canal will be found in the lower second molar, but generally, the canals of the second and third lower molars are similar to those of the first. The pulp chambers are usually smaller, and oftener irregular in outline. The lower third molar has, occasionally, a very large pulp chamber.

VARIATIONS OF THE FORM OF PULP CHAMBERS.

183. Many variations of form occur in the pulp chambers and root canals. The roots of the teeth may be abnormally crooked ; and then the canals will be abnormally crooked. In many instances, the pulp chamber will have in it secondary formations, called nodules, which may be adherent to the walls, or block the mouths of the canals and prevent a broach gliding into them. These also occur occasionally, within the canals, partially blocking the way of the broach. Sometimes the pulp chamber will be filled with nodular deposits so completely that there seems to be no room for the tissues of the pulp. These deposits will have to be removed before the root canals can be reached and entered, after which the canals will generally be found open. These deposits occur within the pulp chambers of any of the teeth ; but they cause annoyance more frequently in the molars.

184. Occasionally lateral openings occur from the root canals to the surface of the root. I have seen more of these from the canals of the lower molars than any other teeth. Generally they follow the course of the dental tubules, and open on the side of the root. They may diverge to one side and curve toward the apex of the root. These cannot often

I

be detected, except in dissections of the root, and occur so rarely they may be ignored in practice.

185.   Sometimes the horns of the pulp approach abnormally near the points of the cusps of some of the teeth, as in the upper first bicuspids, and in the mesio-buccal cusp of the upper first molar.   Then the pulp is more liable to exposure in excavating carious cavities.

PULP CHAMBERS OF THE DECIDUOUS TEETH.

186.   The pulp chambers of the deciduous teeth are proportionally larger, and the thickness of their walls less, than those of the corresponding permanent teeth.   The pulps are, in consequence, exposed with much less penetration of tooth substance, and, therefore, more liable to exposure from caries, or in the use of cutting instruments.   The root canals are generally larger than in the permanent teeth of the same denomination, but are of the same general form.   Also, the same rules for finding the root canals in the permanent molars apply to the deciduous.

ARRANGEMENT OF THE TEETH.

187.   The upper teeth are arranged in the form of a semi-ellipse, the long axis passing between the central incisors (Fig. 126).   In this curve, the cuspids stand a little prominent, giving a fullness to the corners of the mouth.   In different persons there is much variation of the form of the arch within the limits of the normal.   Occasionally the bicuspids and molars form a straight line, instead of a curve, and frequently the third molars are a little outside the line of the ellipse.   The incisors are arranged with their cutting edges forming a continuous curved line from cuspid to cuspid, and this line is continued over the cusps of the cuspids, and the buccal cusps of the bicuspids and molars to the distal surface of the third molars.   From the first bicuspid to the third molar the lingual cusps of these teeth

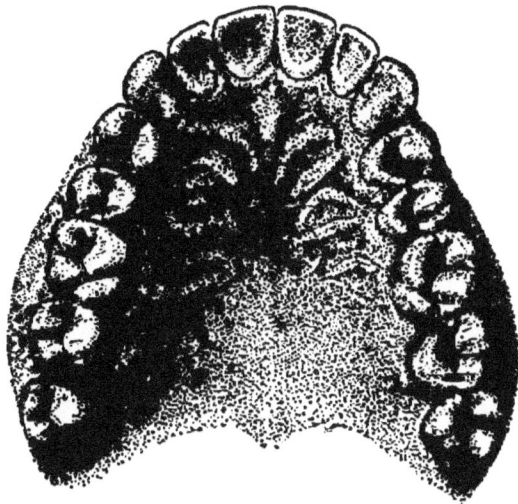

Fig. 126.

Fig. 128.

Fig. 129.

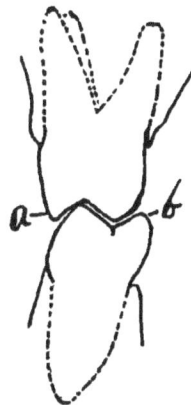

Fig. 130.

FIG. 126* (Par. 187).—ARRANGEMENT OF THE TEETH IN THE ARCH. The arch of the upper jaw.

(Par. 210).—THE TEETH AND THE GUMS AND THE RUGÆ OF THE ROOF OF THE MOUTH.

FIG. 128 (Par. 188).—LABIO-LINGUAL POSITION OF THE INCISORS IN OCCLUSION.

FIG. 129 (Par. 188).—BUCCO-LINGUAL POSITION OF THE BICUSPIDS IN OCCLUSION.

FIG. 130 (Par. 188).—BUCCO-LINGUAL POSITION OF THE MOLARS IN OCCLUSION.

* Illustration, actual size.

form a second line of elevations. Between these two, the lingual and buccal cusps, there is a continuous but irregular valley, or sulcus.

188. The lower teeth are arranged similarly (Fig. 134) but on a slightly smaller curve, so that in occlusion the upper teeth project a little to the labial and buccal of the lower at all points of the arch (Fig. 127). The incisors and cuspids occlude so that the cutting edges of the lower incisors and cusps of the cuspids make contact with the lingual surfaces of the similar teeth of the upper jaw near their cutting edges (Fig. 128). The broad cusped occluding surfaces of the bicuspids and molars of the opposing dentures rest on each other in such a way that the lingual cusps of the upper teeth fit with more or less accuracy into the general sulcus formed by the buccal and lingual cusps of the lower teeth. The buccal row of cusps of the lower teeth, in a similar way, are fitted into the sulcus formed by the buccal and lingual cusp of the upper teeth (Figs. 129 and 130). This arrangement is such that when the teeth are in occlusion, it leaves the buccal inclines of the buccal cusps of the upper teeth outside the buccal surface of the lower teeth (a). And, also, leaves a ledge formed by the abrupt lingual inclines of the lingual cusps of the lower teeth along the lingual line of the occlusion (b). This brings the occluding surfaces of the teeth in the best form of apposition for the purposes of mastication. The forms presented to the cheek and to the tongue hold these soft tissues a little apart from the actual contact points of the occlusion, and thus prevents them from being caught and pinched, or crushed, between the teeth in act of mastication. In youth, while the permanent teeth are taking their places, and before the cusps are properly fitted to the sulci, we often find the cheeks or tongue wounded by being caught between false contact points. With the after movements of the teeth by which they are more perfectly arranged, this difficulty disappears.

189. The line from before backward on which the occlusion occurs is not quite a plain; in the lower jaw it presents a slight curve, or concavity, and in the upper jaw a convexity (Fig. 127, c to d). This concavity of the line of the occluding surfaces of the lower teeth is a little greater than the convexity of the upper, so that the cutting edges of the lower incisors pass a little beyond, and to the lingual of the cutting edges of the upper incisors.

190. In the occlusion, the relative mesio-distal position of the particular teeth of the upper jaw to the lower is important (Fig. 127). At their cutting edges, the upper central incisors are about one-third wider from mesial to distal than the lower centrals. The upper central, therefore, occludes with the lower central, and also with from one-third to one-half of the lower lateral incisor. The upper lateral occludes with the remaining portion of the lower lateral, and the mesial portion of the lower cuspid. The upper cuspid is usually rather broader from mesial to distal than the lower, and in occlusion covers its distal two-thirds and about half of the lower first bicuspid so that its lingual, or triangular ridge, is between two cusps of the lower cuspid and the buccal cusp of the lower first bicuspid, the point of its cusp overlapping the lower teeth. The buccal cusp of the lower first bicuspid occludes in the space between the upper cuspid and upper first bicuspid. This order is now maintained between the bicuspids. The buccal cusp of the upper first bicuspid overlaps (to the buccal) the space between the two lower bicuspids, and its lingual cusp occludes in the sulcus between them, while the buccal cusp of the lower second bicuspid occludes in the sulcus between the two upper bicuspids. The cusps of the upper second bicuspid occlude between the lower second bicuspid and lower first molar. The broad surfaces of the molars come together, so that the mesial two-thirds of the upper first molar covers the distal two-thirds of the lower first molar; and the distal third of

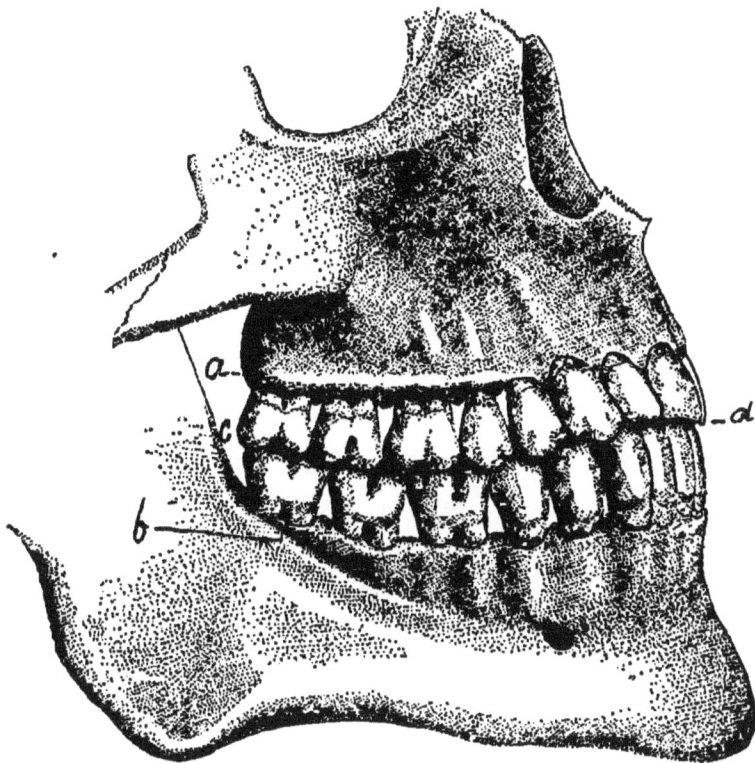

Fig. 127.—Actual size.

FIG. 127 (Par. 188).—ARRANGEMENT OF THE TEETH.   Labial and buccal aspect of the upper and lower teeth as arranged in the arch.

(Par. 197).—THE ALVEOLAR PROCESS OF THE UPPER AND LOWER JAWS, WITH THE TEETH IN POSITION.

the upper first molar covers the mesial third of the lower second molar. This brings the transverse ridge of the upper molar between these two lower teeth. This order is continued between the remaining molars, but less perfectly as the teeth are more irregularly formed. The upper third molar is usually smaller than the lower third molar, yet it generally extends over its distal surface.

191. The long axis of the upper incisors and cuspids are so arranged that their crowns are inclined more or less forward from the perpendicular position, or toward the lip, and slightly toward the median line. The mesial inclination is continued in the bicuspids and molars, diminishing from before backward, and is usually lost at the second or third molar. As a rule, the bicuspids and molars of the upper jaw are also slightly inclined toward the cheek, but in many dentures this inclination is slight, or wanting in the bicuspids and first molars to re-appear in the second and third molars, though it may be absent even in these without necessary malformation.

192. The lower incisors and cuspids are also inclined with their crowns toward the lip, but in less degree than the upper. And even the perpendicular position of these is not inconsistent with a normal arrangement. They have, however, a mesial inclination, but usually much less than the corresponding upper teeth. The lower bicuspids, within the limits of the normal arrangement, vary considerably in their inclinations. Sometimes they have a strong mesial inclination, and at other times they are nearly or quite perpendicular. In many dentures, they also have a lingual inclination, but may be perpendicular or even have a slight buccal inclination. The lower molars usually have a slight mesial and lingual inclination (Fig. 134). In many examples, however, the mesial inclination is wanting, especially in the second and third molars.

193. All the teeth are a little broader from mesial to

distal at or near the occluding surfaces than at their necks.
Therefore, when arranged in the arch with their proximate
surfaces in contact, there is a considerable space between
their necks (Fig. 127). These are known as the inter-proximate, or **V**-shaped, spaces. The sharp angle or apex of the
V-form is toward the occluding surface, or at the contact
point of the proximation, and the open end or base is at the
crest of the alveolar process. In normal conditions this space
is filled by the soft tissues, or gums (136). The average arch
measures about 127 millimeters (5 inches) from the distal surface of the right third molar to the distal surface of the left third
molar, following the curve of the arch. This represents the
average mesio-distal measurement of the crowns of the teeth
of the upper jaw taken collectively. The average measurement of the teeth at their necks is about 89 millimeters (3.5
inches). The remaining 38 millimeters (1.5 inches) represent the average sum of the inter-proximate spaces taken
collectively.

194. On account of differences in the conformation of
the crowns and the inclination of the teeth, the inter-proximate spaces vary much in width in different dentures. They
are much wider between bell-crowned teeth than between
thick-necked teeth ; but some inter-proximate space exists in
every normal denture. When the crowns of the incisors and
cuspids are much inclined toward the lip, the necks of the
teeth form a smaller circle than the line of the contact points
of the proximation, and in this way the inter-proximate
spaces may be considerably narrowed. Generally, the inter-
proximate space is wide between the necks of the central
incisors. The suture joining the maxillary bones passes between the roots of these teeth, and they are somewhat farther
apart than the roots of central and lateral incisors, or those
of the lateral incisor and the cuspid. Therefore, in these
latter, the inter-proximate spaces are of less width. Between
the bicuspids the inter-proximate spaces are wider at the

necks of the teeth than between the anterior teeth, on account
of the proportionally broader crowns. The widest inter-proximate spaces are usually between the necks of the molars.

195. The points of the proximate contact in the best
formed arches are very near the occluding surfaces of the
teeth. In imperfectly developed teeth, in which the crowns
are much rounded toward the occluding surfaces, the contact point is more toward the gingival. In the incisors and
cuspids it is in direct line with the cutting edges. In the
bicuspids the contact is with the buccal angles and in line
with buccal cusps. The mesial and distal flattened surfaces
of these teeth converge to the lingual to such an extent that,
though they are arranged in arch form, the contact points remain on the buccal angles. In many excellent dentures
there is a decided inter-proximate space opening to the lingual, but in thick-necked teeth and those of a more rounded
contour, the contact points are often more toward the lingual,
and there is no appreciable lingual inter-proximate space. In
the molars the contact points as a rule are removed rather
more to the lingual, but still in the best formed dentures
they will be found nearly in line with the buccal cusps.
Between the upper first and second molars, the contact point
is often extended toward the lingual by the prominent disto-
lingual cusp of the first molar; and, even when otherwise,
the general rounding of the distal surfaces of the upper
molars often brings the contact point near the middle line of
the teeth. In lower first molars the large distal cusp brings
the contact point with the second molar close to the buccal
side, with a considerable lingual inter-proximate space. If
the distal cusp is small the contact point is usually extended
toward the lingual, often as far as half the labio-lingual
breadth of the teeth. Between the second and third molars
the contact point is most frequently near the central line of
the teeth. In the best formed dentures the form of the
proximate contact is such as to prevent food from being

crowded between the teeth in mastication; and, therefore, such as to keep these spaces clean and the inter-proximate gingivus in health.   But many faulty forms are met with which allow food to leak through into the space and crowd the gum away, forming a pocket for the lodgment of débris, giving opportunity for decomposition, and resulting in caries of the proximate surfaces, or disease of the gum and peridental membrane. Exceptionally, cases are met with in which the teeth stand so widely apart that the spaces are self-cleaning.   The form of the inter-proximate spaces is very variable.   It is best studied in skulls in which the teeth are all present, and by careful consideration of the forms of the proximate surfaces of the teeth, together with their relative positions.

THE ALVEOLAR PROCESS AND ALVEOLI.

196.   The alveolar process is the projecting portion of the maxillary bones within which the roots of the teeth are lodged in alveoli, or sockets, accurately fitted to their surfaces (Figs. 131 and 132).   The form of the alveolar process seems to depend on the teeth, the conformation of their roots, and their arrangement in the arch.   If any teeth are misplaced, or from any cause stand out of the regular and normal line, the alveolar process is formed about their roots in this irregular position.   Also, when teeth are lost, the alveolar process mostly disappears by absorption, and the remaining portions of the alveoli are filled with bone.

197.   Normally, the alveolar process envelops the roots of the teeth to within a short distance of the gingival line (Figs. 127 and 133), varying from one to three millimeters in the young adult.   This distance increases somewhat with increasing age.   The margins of the process are reduced to a thin edge about the necks of the teeth on both the labial and lingual sides of the incisors and cuspids of the upper jaw.   About the lingual sides of the necks of the bicuspids

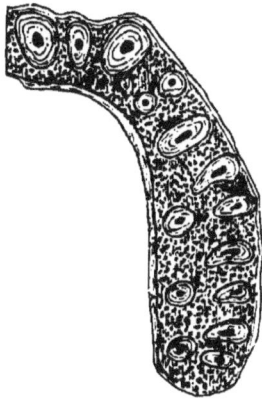 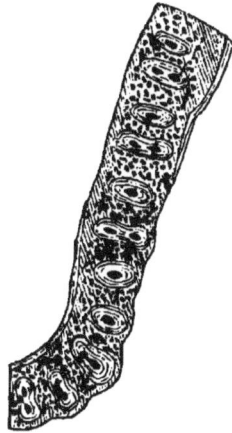

Fig. 131.   Fig. 132.

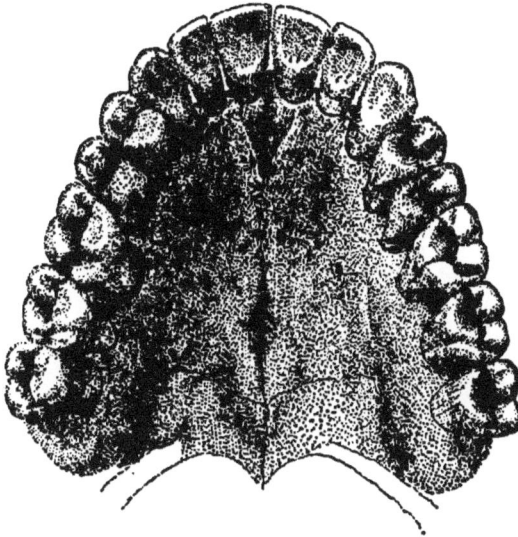

Fig. 133.

FIG. 131 * (Par. 196).—SECTION OF THE ALVEOLAR PROCESS AND ROOTS OF THE TEETH OF THE UPPER JAW, showing the roots of the teeth in position.

FIG. 132 * (Par. 196).—HORIZONTAL SECTION OF THE ALVEOLAR PROCESS AND ROOTS OF THE TEETH OF THE LOWER JAW, showing the roots of the teeth in position.

FIG. 133* (Par. 197).—THE TEETH AND LINGUAL PORTION OF THE ALVEOLAR PROCESS OF THE UPPER JAW, AND THE BONES OF THE ROOF OF THE MOUTH.

---

* Illustration, actual size.

and molars the margins are also reduced to a thin edge, becoming slightly thickened about the second and third molars, especially of the latter. On the buccal sides of these, a thickening of the immediate margin of the process, in the form of a marked ridge, begins about the first or second bicuspid, more commonly between these two, and extends to the distal of the third molar (Fig. 127, *a*). This ridge varies in different examples, from a very slight thickening of the immediate margin, to a thickness of two or three millimeters. It forms a margin standing squarely out from the necks of the teeth. The process then thins away so that, in many instances, the buccal roots of the teeth, especially the mesial root of the first molar, have but a thin covering of bone.

198. Anteriorly, the bony covering of the roots of the incisors presents much variety. In some examples, the middle portion of the roots has but a slight covering of bone, but more generally it is progressively thickened from the neck to the apex. The roots of the cuspids are prominent toward the lip, and, for most of their length, have only a thin bony covering, and this forms a ridge along the line of the root, which may easily be traced with the finger through the soft tissues of both the gum and lip. In many instances, the bony covering is entirely wanting for a little space near the middle of the length of the root of the cuspid, the buccal root of the first bicuspid, the mesial root of the first molar, and, occasionally, of other teeth.

199. On the lingual side of the upper teeth (Fig. 133), the progressive thickening of the alveolar process, from the gingivus toward the apex of the root, is much greater; so that the roots of the teeth seem to lie toward the labial and buccal side of the alveolar process (Fig. 131). Even the large lingual root of the first upper molar, diverging strongly to the lingual, seldom forms a ridge or prominence of the process covering its lingual surface.

200. The anterior palatine foramen is in the median

line, just behind the central incisors (Fig. 133). It is funnel-shaped with a broad opening to the palatine surface of the bone. The bone is often quite thin between this and the roots of the central incisors. The posterior palatine artery runs in a deep groove in the surface of the bone, very close to the apex of the lingual roots of the upper third, and the upper second molar. This artery is occasionally wounded, or cut, in efforts to extract the roots of these teeth.

201. In the lower jaw, the immediate margins of the alveolar process, on the labial sides of the incisors, are rather thicker than in the upper jaw, often amounting to a decided ridge. This, however, thins away rapidly, so that the middle portion of the roots has but a thin covering of bone. At the cuspid, the margin of the process is very thin, but a gingival ridge, corresponding with that of the upper jaw, though not so prominent, begins at about the first or second bicuspid and runs to the second molar. This thins away over middle of the length of the roots of the bicuspids and first molar. At the second lower molar, the rising of the external oblique ridge for the formation of the anterior border of the coronoid process (Fig. 127, b), causes a thickening of the buccal bony covering of the root, while at the third molar this ridge rises to a level with the gingival margin of the process, making the bony covering on the buccal side of the root about one-fourth of an inch thick (Fig. 132). This is of importance with reference to the extraction of the roots of this tooth. Indeed, the second and third molars of the lower jaw are fixed in alveoli, hollowed out in the lingual side of the body of the bone rather than in a process or ridge on the bone, as with the teeth anterior.

202. On the lingual side of the lower anterior teeth, the immediate gingival border of the alveolar process is a smooth thin edge, and the covering of bone over the roots progressively thickens toward their apexes. In this portion, the process is high, and the labio-lingual thickness is only

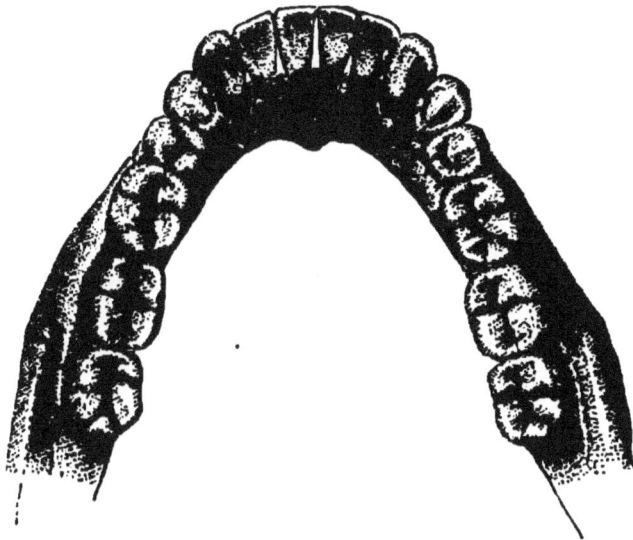

Fig. 134.—Actual size.

Fig. 136.—Actual size.

FIG. 134 (Par. 202).—LOWER TEETH AND MAXILLA, as seen from above.

FIG. 136 (Par 210) —THE LABIAL AND BUCCAL ASPECT OF THE TEETH AND GUMS OF THE UPPER JAW.

K

sufficient to envelop the roots of the teeth and give them support (Figs. 134 and 135). From the first bicuspid backward the height of the process diminishes rapidly, and the line of the dental arch, and of the alveoli of the teeth, passes diagonally across the line of the curve of the body of the bone from the buccal side at the first bicuspid to the lingual side at the third molar, and sinks into the body of the bone (Figs. 134 and 135). The immediate gingival border remains thin as far as the first or second molar, but the bony covering of the roots is rapidly thickened toward their apexes.

203. This is caused largely by the thickening of the body of the bone, produced by the rise of the mylohyoid ridge (Fig. 135, a), which begins in front below the apexes of the roots and rises progressively backward nearly to the gingival border of the alveolus at the third molar. The greatest thickness is over the lingual side of the apical half of the roots of the second molar, and of the crossing of the lingual side of the coronal half of the roots of the third molar. At the third molar this ridge is thinner, and at the distal angle often very slight, while the lingual covering of the roots of the tooth may be very thin below this ridge. Therefore, it is usually easy, when necessary in extracting, to force the roots of the lower third molar to the lingual and distal with a lever suitably arranged against the crown of the second molar, as with the Physic's forcep, or an elevator.

204. The gingival margin of the septi of the alveoli of the roots of the anterior teeth are rounded from labial to lingual, but this diminishes rapidly from the cuspid backward. Between the molars it is a straight line, or presents but a slight concavity, so that the highest points of the gingival border of the process are on the buccal and lingual surfaces, or at the angles, of the necks of these teeth, which is important in fitting bands of crowns or any similar processes.

205. The alveolar process is composed of an outer and an inner plate of moderately compact bone, and between these, very open cancellous or spongy bone; so that in young persons the process may be forced more or less to one side, or bent out of position, without definite fracture. The outer compact plate forms the outer surface of the bone, and the inner plates line the alveoli of the teeth. These latter are very thin, and supported on all sides by the cancelous structure. In the lower jaw the substance of the bone is more compact and stronger than in the upper, especially about the molar teeth, where the alveoli are in the substance of the body of the bone.

THE PERIDENTAL MEMBRANE.

206. The peridental membrane invests the roots of the teeth from the gingival line to the apexes of the roots like a sack. It lines every part of the alveoli, and, passing over the gingival margins, is continuous with the periosteum and gums covering the outer plates of the alveolar processes. It is one membrane attached on one side to the root of the tooth, and on the other to the inner wall of the alveolus. It is composed of connective tissue which supports an abundant supply of blood vessels, nerves and lymphatics. With these are intermingled strong fibers of white fibrous tissue which pass from the cementum of the root of the tooth to the bony walls of the alveolus. The ends of these are built firmly into each, forming a strong attachment of the root to its alveolus. In childhood and youth, this membrane is comparatively thick, and allows considerable motion of the tooth in its socket. As age advances, it becomes thinner and the motions of the teeth are more restricted. A bundle of nerves and one or more arteries enter the alveolus near the apex of the root (the apical space), and, subdividing, several arterial twigs and nerve bundles pass toward the gingivus; while others enter the apical foramen and pass to the pulp

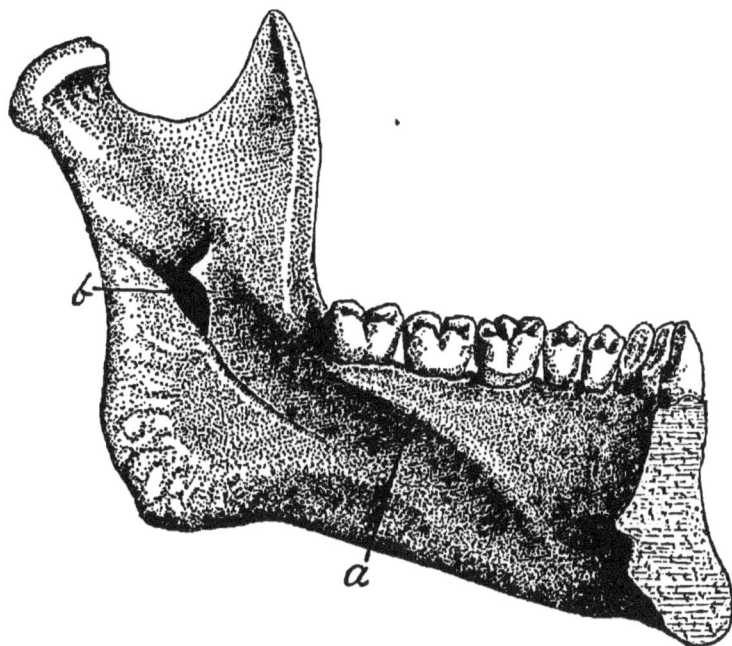

Fig. 135.—Actual size.

FIG. 135 (Par. 202).—LINGUAL SURFACE OF THE LEFT HALF OF THE LOWER MAXILLA AND TEETH. a, Mylo-hyoid ridge; b, inferior dental foramen.

of the tooth. At the gingivus, the blood-vessels become continuous with those of the periosteum and gums.*

THE GUMS.

207. The gums are the soft tissues covering the alveolar processes and investing the necks of the teeth. They are composed of connective tissue containing many white fibres intermingled, forming a firm membranous mass, continuous with the periosteum beneath and peridental membrane at the necks of the teeth. They are covered with a strong outer coat of squamous epithelium. They are richly supplied with blood-vessels and nerves ; but, under normal conditions, they are not very sensitive to pressure or moderate violence, but, in pathological conditions they may become very sensitive.

208. On the labial and buccal side of the alveolus, the gums (Fig. 135) are thin and firm near the necks of the teeth and strongly attached to the periosteum. As they pass from the necks of the teeth toward the base of the alveolar ridge they become softer and loose their attachment to the periosteum, and become merged into the buccal mucous membrane of the lips and cheeks. In the passage from the gums to these mucus surfaces several loose folds are formed, the most notable being the frenum labium of the upper jaw at the median line, passing from near the inter-proximate space of the central incisors to the median line of the upper lip. A similar, though less prominent fold connects the median line of the lower lip with the gums. Occasionally, folds of less prominence are seen in the neighborhood of the bicuspids and first molars.

209. On the lingual side of the arch in the lower jaw, the conditions are much the same. The gums are firmly

---

* For the details of this subject the student is referred to the author's treatise on "The Histological Character of the Periosteum and Peridental Membrane."

adherent to the periosteum near the teeth, but lower down are merged into the mucous membrane of the floor of the mouth. The median line of the tongue is connected with the median line of the gums by a strong fold, known as the frenum linguæ. As an abnormal congenital condition this is sometimes connected so near the tip of the tongue as to prevent its protrusion over the teeth, constituting the condition known as "tongue-tied." This is generally corrected by the muscular efforts of the tongue, which stretch the membrane sufficiently to accommodate its motions.

210. On the lingual side, in the upper jaw, the gums (Fig. 136) are usually of greater extent and thicker. They cover the entire roof of the mouth, to the conjunction of the hard and soft palate, as a hard dense layer. In the anterior portion, a series of irregular ridges, known as the rugæ, radiate from the median line toward both sides, stopping short of the gingivæ. In a large collection of examples almost innumerable forms of these rugæ may be noted.

211. The gingivæ, or gingivus, is that portion of the gum tissue investing the neck of the tooth crownwise from the attachment of the peridental membrane at the gingival line. It is also termed the free margin of the gum. The length of the gingivæ, from the attachment to the neck of the tooth, varies in different teeth in the same mouth, and in different adults, from about one to about four millimeters. It is often much greater in young persons, but the length usually diminishes as age advances. The free margin fits around the neck of the tooth closely; but a thin, flat instrument is easily passed between it and the tooth to the attachment at the gingival line. As the gingival line of the tooth is at the border of the enamel, or, at the junction of the enamel and cementum, it follows that the gingivæ inclose the immediate border of the enamel, and cover this part of the crown of the tooth. In young persons, we often see one-half of the length of the crowns of the teeth buried in the gingivæ, even

after the teeth are regarded as fully in place. As age advances, the gingivæ become shorter, showing more of the
own, and finally recede to very near the gingival line.

212. The gingivæ also fill the inter-proximate spaces in
the form of septi passing between the teeth from labial, or
buccal, to lingual. This portion of the gingivus is much
longer than that on the labial and lingual surfaces of the
teeth (Fig. 136). In normal conditions it reaches from the
gingival line to the contact point between the teeth, completely filling the space and preventing accumulation of
débris. The form presented by the gingivæ on the labial
and buccal surfaces, is a series of imperfect semi-circles with
the concavity toward the occluding surfaces of the teeth, and
with the points of junction of these extending into, and filling the inter-proximate spaces (Fig. 136). On the lingual
side of the arch, the conditions are much the same, but the
points of the gingivæ between the teeth are less prominent
(Fig. 126). As age advances and the gingivæ recede toward,
or even to, the gingival line, the septi of soft tissue sometimes fail to fill the inter-proximate spaces. This may induce pathological conditions by affording space for lodgment
of débris in pockets, which favors fermentation. A like
condition is also induced frequently by a faulty form of
proximate contact, which allows food to be forced into. the
inter-proximate space, and break down the gum septum by its
pressure.

INDEX.

L